oks are to be returnd

D1643519

Current Topics in Microbiology
233/I and Immunology

Editors

R.W. Compans, Atlanta/Georgia
M. Cooper, Birmingham/Alabama
J.M. Hogle, Boston/Massachusetts · Y. Ito, Kyoto
H. Koprowski, Philadelphia/Pennsylvania · F. Melchers, Basel
M. Oldstone, La Jolla/California · S. Olsnes, Oslo
M. Potter, Bethesda/Maryland · H. Saedler, Cologne
P.K. Vogt, La Jolla/California · H. Wagner, Munich

Springer
Berlin
Heidelberg
New York
Barcelona
Budapest
Hong Kong
London
Milan
Paris
Santa Clara
Singapore
Tokyo

Reoviruses I

Structure, Proteins, and Genetics

Edited by K.L. Tyler and M.B.A. Oldstone

With 29 Figures, 3 in Color and 13 Tables

 Springer

Kenneth L. Tyler M.D.
University of Colorado
Health Sciences Center
Department of Neurology
Campus Box B-182
4200 East 9th Avenue
Denver, CO 80262
USA

Michael B.A. Oldstone, M.D.
The Scripps Research Institute
Department of Neuropharmacology
Division of Virology
10550 N. Torrey Pines Road
La Jolla, CA 92037
USA

Cover Illustration: Surface-shaded views of the reovirus T1L virion (top), ISVP (middle), and core particle (bottom) based on three-dimensional image reconstruction of data obtained by cryoelectron microscopy. (See Figure 3, Chapter 1 for additional details. Images provided by Dr. M.L. Nibert).

Cover Design: design & production GmbH, Heidelberg

ISSN 0070-217X

ISBN 3-540-63946-2 Springer-Verlag Berlin Heidelberg New York

This work is subject to copyright. All rights are reserved, whether the whole or part of the material is concerned, specifically the rights of translation, reprinting, reuse of illustrations, recitation, broadcasting, reproduction on microfilm or in any other way, and storage in data banks. Duplication of this publication or parts thereof is permitted only under the provisions of the German Copyright Law of September 9, 1965, in its current version, and permission for use must always be obtained from Springer-Verlag. Violations are liable for prosecution under the German Copyright Law.

© Springer-Verlag Berlin Heidelberg 1998
Library of Congress Catalog Card Number 15-12910
Printed in Germany

The use of general descriptive names, registered names, trademarks, etc. in this publication does not imply, even in the absence of a specific statement, that such names are exempt from the relevant protective laws and regulations and therefore free for general use.

Product liability: The publishers cannot guarantee the accuracy of any information about dosage and application contained in this book. In every individual case the user must check such information by consulting other relevant literature.

Typesetting: Scientific Publishing Services (P) Ltd, Madras

SPIN: 10559726   27/3020 – 5 4 3 2 1 0 – Printed on acid-free paper

This book is dedicated to the memory of Bernard Fields. KLT would also like to thank his wife Lisa and children Max and Eric for their encouragement and support.

Foreword

Mammalian reoviruses do not cause significant human disease, yet they have commanded considerable interest among virologists, geneticists, and biochemists. Their genome consists of ten segments of double-stranded RNA and, because a number of reovirus strains exist, it has been possible to isolate and study the function and structure of a single gene of one strain placed upon the backbone of the nine other genes of the second strain using genetic reassortant techniques. Such manipulations have led to fundamental studies on binding, entry, replication, transcription, assembly, and release. In addition, reovirus type 1, compared to type 3, displays different disease phenotypes in the mouse. Reassorting genes between reovirus 1 and 3 allows a dissection of several important questions concerning host–virus interactions and understanding the molecular basis of the associated disease. Most of the major players who have made seminal contributions in this area have contributed chapters to these two volumes of Current Topics in Microbiology and Immunology 233. One whose presence is found throughout both volumes, but is not an active contributor, is the late Bernard Fields, who died on 31 January 1995.

Bernie Fields had a profound influence not only in the reovirus field, but in the arena of virology as a whole, especially viral pathogenesis. It was his appreciation of colleagues in the reovirus field and their appreciation of him, coupled with his strong commitment to the training of independent scientists, that made these two volumes dedicated to him a labor of love. Fields' significant research in this area is matched by the accomplishments of a number of committed and productive biomedical scientists that he trained, many of whom have contributed to these two volumes. It is these scientific children who will keep Fields' memory alive. Specific thanks is given to Ken Tyler, a former student and associate of Bernard Fields who was primarily responsible for organizing and collecting papers for these two volumes.

Not only do these two volumes stand as a testimony to the respect Bernard Fields earned and received from others, but the

authors have unanimously agreed that their honorarium received
from Springer-Verlag for the publication of these two volumes
will be used to provide lecture series on microbial (viral) pathogenesis to be given in honor of Bernie.

La Jolla, California   MICHAEL B. A. OLDSTONE
March, 1998

Preface

There were two common and interlocking themes that guided the selection of contributors to these two volumes, namely, Dr. Bernard N. Fields and reoviruses. Dr. Fields' untimely death at the age of 56 on 31 January 1995 was a tremendous loss to his family, to his friends and colleagues, to the field of virology in particular, and to science in general. All of the contributors to these volumes had close ties to Bernie, and all of us still mourn his loss. I suspect that they, as I, frequently still find themselves starting to pick up the telephone to tell Bernie about an exciting new finding or to get his advice about a perplexing scientific, professional, or personal problem. Bernie was always available for these phone calls, and was genuinely proud and excited about what his former students and trainees and his colleagues were accomplishing.

Many of the contributors to these two volumes trained as postdoctoral fellows in Dr. Fields' laboratory, including Drs. Nibert, Coombs, Ramig, Schiff, Brown, Dermody, Tyler, Sherry, Rubin, and Virgin (listed in the order in which their chapters appear). Others were collaborators, close colleagues, and friends (Drs. Shatkin, Joklik, Lee, Greene, Jacobs, Maratos-Flier, and Samuel). In many cases these individuals have enlisted colleagues and trainees from their own laboratories as coauthors. Bernie was extremely proud of the continuing multigenerational expansion in the reovirus "family", and the extraordinarily high quality of work that was being performed by so many talented researchers.

Difficult choices must inevitably be made in selecting contributors to a work of this type, and as a result inadvertent omissions occur. For these the editors take full responsibility, and offer apology in advance to any who may feel slighted. It is undoubtedly a further tribute to Bernie that everyone invited to contribute to these volumes accepted immediately and enthusiastically. It is sad, though (and the bane of all editors) that not all those who accepted were ultimately able to contribute. The editors and contributors have agreed to forego any royalties in conjunction with these volumes and instead to utilize these funds to endow a Fields' Lectureship in Microbial Pathogenesis to be held in conjunction with the FASEB meetings.

Perhaps the hardest editorial decision, required for reasons of space and logical coherence, was to restrict contributors to those still actively involved in reovirus research. Many of Bernie's former trainees who were excluded in this way have made eminent contributions in other branches of virology, immunology, and molecular biology or in other fields of science. I can only hope that their voices will be heard in future volumes dedicated to Bernie's memory.

Reoviruses have been and continue to be an important viral system for understanding the molecular and genetic basis of viral pathogenesis, a theme that was central to much of Bernie's own laboratory research. Bernie clearly recognized that understanding basic aspects of the structure, molecular biology, and replication strategy of viruses is critical to developing a complete and accurate picture of pathogenesis. From a personal perspective, I always remember Bernie's infectious excitement when he thought that some new research finding or observation helped link some fundamental aspect of basic virology with an improved understanding of how viruses ultimately cause disease. In this spirit, the contributions to these volumes run the gamut from studies in basic reovirology to the use of reoviruses to explore pathogenesis in vivo.

Chapters in Volume I deal with fundamental reovirology, including studies of virion structure, the structure and function of individual viral structural proteins, and the nature of the virus genome and its assembly. Also included in the first volume are chapters dealing with temperature-sensitive (ts) mutants, and the effects of reovirus interaction with cell surface receptors.

In Volume II the focus shifts to emphasize the effects of reoviruses' interaction with target cells and of reovirus infection on individual organ systems. This second volume includes chapters dealing with the molecular mechanisms of reovirus persistent infections and reovirus-induced apoptosis. These chapters are followed by individual chapters dealing with reovirus infection of particular organs including the heart, liver, biliary, endocrine, and nervous systems. These selections are not intended as an exhaustive catalogue of pathology but rather to highlight reoviruses' diverse effects on many biological systems in vivo and the mechanisms by which these occur. Finally, as Bernie clearly recognized, it is impossible to truly understand viral pathogenesis without also understanding the role played during infection of the various components of the host's immune system. The concluding chapters of the second volume deal with selected aspects of reoviruses and their interaction with cytokines, antibodies, and the cellular immune system.

These two volumes should be considered a selective snapshot of the current state of the art in many areas of reovirus research. All such collections suffer from innate biases, with some topics being overemphasized and others inadvertently omitted. For all these imperfections the editors take full responsibility. We can only hope that the readers of these volumes will find that the pleasure obtained from seeing so much good work being carried out by so many gifted people will outweigh the annoyances engendered by any deficiencies. We can also only hope that Bernie, were he still alive, would have felt the same way and considered this a sort of Festschrift, celebrating the profound influence he had on all of us.

Denver, Colorado   KENNETH L. TYLER
March 1998

List of Contents

List of Contents
of Companion Volume 233/II

List of Contributors

(Their addresses can be found at the beginning of their respective chapters.)

LIVERPOOL
JOHN MOORES UNIVERSITY
AVRIL ROBARTS LRC
TEL. 0151 231 4022

Structure of Mammalian Orthoreovirus Particles

M.L. Nibert

Institute for Molecular Virology, The Graduate School, University of Wisconsin-Madison, Madison, WI 53706, USA
Department of Biochemistry, College of Agricultural and Life Sciences, University of Wisconsin-Madison, Madison, WI 53706, USA

1 Introduction to Mammalian Orthoreoviruses

An overview of the mammalian orthoreoviruses and comparisons with other double-stranded RNA (dsRNA) viruses are included to stimulate speculation about the evolution of these viruses and their structures.

1.1 Similarities Among Isolates in the Three Serotypes

The orthoreoviruses comprise one genus in the *Reoviridae* family (Fig. 1). The prototype members are a group of viruses that infect mammals and are termed the mammalian orthoreoviruses. These viruses are closely related in sequence and can exchange their ten genomic dsRNA segments by reassortment both in the laboratory and in nature. Type 1 Lang (T1L), type 2 Jones (T2J), and type 3 Dearing (T3D) are the most commonly studied of the mammalian isolates. There are few restrictions to infections of different mammals by these viruses; thus strictly human, murine, etc., strains of mammalian orthoreoviruses are not recognized. Although there are three serotypes (types 1, 2, and 3 in the names above), this designation may well reflect only the nature of the S1 gene segment in a particular isolate, in that S1 encodes the major determinant for neutralizing antibodies (WEINER and FIELDS 1977), the cell attachment protein σ1 (LEE et al. 1981). Where sequences

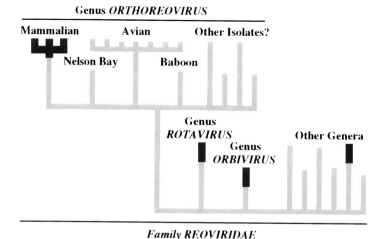

Fig. 1. Possible phylogenetic tree for the genus *Orthoreovirus* and the family *Reoviridae*. The mammalian and avian groups of orthoreovirus isolates are shown in addition to two isolates of less certain affiliation: Nelson Bay virus and baboon reovirus. *Gray lines* show the uncertain degree of relationship among the different groups of orthoreoviruses. *Black lines* indicate that sequences have been ascertained for multiple genes from several isolates, permitting the degree of relationship among these isolates to be determined. Isolates from genera other than *Orthoreovirus* are shown in less detail. Although not accurately depicted by the lengths of lines here, it is obviously expected that greater relationship exists between isolates within each genus than between isolates in different genera

have been determined from multiple isolates, the other gene segments show little or no serotype-specific variation, reflecting the frequent mixing of these other segments with each of the serotype-specific S1 segments during reassortment in nature (CHAPPELL et al. 1994; KEDL et al. 1995; GORAL et al. 1996). Sequence alignments reveal that the σ1 proteins from isolates of different serotypes require gaps in their alignments and are conserved to as little as 26% (DUNCAN et al. 1990; NIBERT et al. 1990). In contrast, the proteins encoded by the other gene segments are colinear in alignments and conserved to at least 86% in isolates of different serotypes (WIENER and JOKLIK 1989; CHAPPELL et al. 1994; KEDL et al. 1995; GORAL et al. 1996). As a result, the structural details described for one mammalian orthoreovirus are thought to be essentially true for all others, except for some differences among the σ1 proteins of the three serotypes.

1.2 Divergence from Other Orthoreoviruses

A large group of isolates from birds, the avian orthoreoviruses, appear related to the mammalian isolates in the protein-coding pattern of their ten dsRNA segments, the general organization of proteins within virions (NI et al. 1993; DUNCAN 1996), and minor levels of shared antigenicity. While the exchange of gene segments is expected to be possible among most of these avian viruses, it has not been demonstrated between mammalian and avian isolates, suggesting that these groups are genetically divergent and constitute distinct quasispecies of orthoreoviruses (Fig. 1). Such divergence is indeed indicated by the recently reported sequences of S1 gene segments from four avian isolates (e.g., LIU and GIAMBRONE 1997). Several other isolates that appear to be orthoreoviruses based on the protein-coding pattern of their ten gene segments, particle morphology, and antigenic properties appear to be distinct from the major mammalian and avian groups. These include isolates from a flying fox (Nelson Bay virus; WILCOX and COMPANS 1982), baboon (DUNCAN et al. 1995), python (AHNE et al. 1987), and rattlesnake (VIELER et al. 1994). Definition of the evolutionary relationship between the classical mammalian orthoreoviruses and the avian and other isolates awaits sequencing of the latter's genomes. Similarly, enough is not yet known to permit an informative comparison of the subunit structures of the mammalian and other orthoreoviruses; nonetheless, as additional data become available, such comparisons may be useful for understanding structure–function relationships within the viral particles. The remainder of this review will focus on the classical mammalian isolates and will refer to these simply as reoviruses.

1.3 Divergence from Other Viruses in the *Reoviridae* Family

Sequence data indicate that the reoviruses are highly divergent from members of the other eight genera in the *Reoviridae* family (Fig. 1), including two of three whose members infect mammals, i.e., rotaviruses and orbiviruses. In fact, it is sometimes difficult to discern which proteins from viruses in the different genera are analogous to each other, let alone to identify regions of homologous sequences. Some basic structural similarities among these viruses can be defined:

1. Virions lack a lipid envelope.
2. Proteins in virions are arranged in several concentric layers (two or three depending on how they are defined).
3. dsRNA genome segments, ten to 12 in number, are condensed in the particle center and have no protein contacts over much of their lengths.
4. Outer protein layer(s)

   a) Subunits of either one or two major proteins are arranged on a T = 13(laevo) icosahedral lattice.
   b) The arrangement of these subunits creates radially directed channels at the penta-coordinated (P1) and hexa-coordinated (P2 and P3) positions in the lattice (132 total).
   c) One or more additional protein may decorate the lattice, commonly by binding within one class of channel.
   d) The major protein in the lattice is centered around local threefold axes.

5. Inner protein layer(s)

   a) Subunits of either one or two major proteins, at least one of which is present in 120 copies per virion, are arranged on a T = 1 icosahedral lattice.
   b) The arrangement of these subunits creates a radially directed channel at each penta-coordinated position in the lattice.
   c) Additional proteins decorate the lattice, at least one by binding within the penta-coordinated channel and at least one of which is the viral RNA polymerase.

Proteins with the following activities are present:

-   Binding to cell surface receptors
-   Penetration of a cell membrane bilayer
-   Transcription of the viral mRNAs: RNA polymerase, other possible activities
-   Capping of the viral mRNAs: RNA triphosphatase, RNA guanylyltransferase, RNA 7-N-guanosine methyltransferase, RNA 2′-O-ribose methyltransferase

However, some of these attributes remain tentative, since a structure at a resolution of 30 Å or less has been reported for isolates from only four of the nine genera (reovirus; DRYDEN et al. 1993), rotavirus (YEAGER et al. 1990; SHAW et al. 1993), orbivirus (HEWAT et al. 1994), and aquareovirus (SHAW et al. 1996). Even when analogous protein regions from the viruses in these different genera can be clearly identified, such as the catalytic regions of the RNA polymerase, the divergence is striking (KOONIN 1992). Thus, until more is known about the structures and functions of proteins from these viruses, caution is warranted when proposing detailed analogies among their protein components (e.g., those between reovirus and rotavirus proteins mentioned below).

1.4 Comparison with Other dsRNA Viruses
Outside the *Reoviridae* Family

Several other viruses with dsRNA genomes are classified in other families –
Birnaviridae, *Cystoviridae*, *Totiviridae*, *Partitiviridae*, and *Hypoviridae* – because of
more divergent characteristics (MURPHY 1996). For example, each of these viruses
contains only one to three dsRNA gene segments. Despite such evident divergence,
some structural similarities with the *Reoviridae* are also seen. For example, the
Birnaviridae have only a single icosahedral capsid, but it is organized with T = 13
symmetry (BOTTCHER et al. 1997), comparable to the outer capsids of the *Reoviridae*
except that it may be in the T = 13(dextro) configuration in some cases (OZEL and
GELDERBLOM 1985) as opposed to the T = 13(laevo) configuration seen in the
Reoviridae. The *Totiviridae* also have only a single icosahedral capsid, but it is
organized with a distinctive form of T = 1 symmetry in which a protein present in
120 copies per particle is arranged in asymmetric dimers to create the T = 1 lattice
(CHENG et al. 1994). This pattern is probably similar to the inner capsids of the
Reoviridae (SHAW et al. 1993). Such similarities may represent clues not only about
the evolutionary relationships among the divergent families of dsRNA viruses but
also about some shared functions in their replication cycles that are mediated in
similar fashions by the analogous T = 13 and T = 1 capsids.

2 Organization of Reovirus Particles

The analysis of reovirus structure has been aided by our capacity to obtain not only
large amounts of purified virions but also several types of subvirion particles that
are missing or have alterations in certain of the structural proteins. Infectious
subvirion particles (ISVPs) and cores obtained by protease treatments of virions
have been especially useful, as described below.

2.1 Concentric Protein Layers

The ten genomic dsRNA segments of reoviruses are known to encode 11 proteins,
one of which appears to initiate at either of two in-frame start codons (WIENER
et al. 1989). Three of these 11 proteins are nonstructural in that they play roles
restricted to the infected cell and are not incorporated into the mature, infectious
virions that are released from cells upon lysis. The other eight proteins are packaged into virions, where each is present in a defined number of copies (Table 1) and
occupies characteristic, symmetry-related positions in one of the icosahedrally arranged capsids. The contacts that these proteins are known or suspected to make
with each other, as well as with the dsRNA genome segments, are summarized in
Table 2.

6   M.L. Nibert

Table 1. Properties of proteins in reovirus virions

Protein	Size[a]	Copy number	Location[c]	Processing or cofactors	Presence in subvirion particles
λ1	137,400[d]	120	Inner	Binds Zn^{2+}	ISVP, core
λ2	144,000	60	Outer	–	ISVP, core
λ3	142,300	12	Inner	–	ISVP, core
μ1	76,300	600	Outer	N-myristoylated; cleaved into μ1N (4200) and μ1C (72,100)	ISVP[e]
μ2	83,300	12[f]	Inner	–	ISVP, core
σ1	49,200[g]	36[f]	Outer	–	ISVP[h]
σ2	47,200	120[f]	Inner	–	ISVP, core
σ3	41,200	600	Outer	Binds Zn^{2+}	–

ISVP, infectious subvirion particle.
[a] Based on deduced amino acid sequence, to the nearest 100 Da.
[b] Estimated from image reconstructions and/or by sodium dodecyl sulfate-polyacrylamide gel electrophoresis (SDS-PAGE).
[c] Inner or outer capsid (see Fig. 2).
[d] Probably nearer 142,000: premature termination codon in published sequence (S. Noble et al. unpublished).
[e] Cleaved by exogenous protease: μ1 yields μ1δ (63,000) and φ (13,000); μ1C yields δ (59,000) and φ. Exact site of cleavage depends on protease (NIBERT and FIELDS 1992).
[f] Lowest confidence in these copy numbers.
[g] Differences between serotypes: type 2 Jones σ1, 50,500; type 1 Lang σ1, 51,400 (DUNCAN et al. 1990; NIBERT et al. 1990).
[h] Cleaved within fiber region in some isolates, including T3D (NIBERT et al. 1995).

Table 2. Contacts by reovirus proteins in virions

Protein	Interaction with other protein or dsRNA								
	λ1	λ2	λ3	μ1	μ2	σ1	σ2	σ3	dsRNA
λ1	?	+	+	+	?	–	+	–	+
λ2	+	+	+	+	–	+	–	+	–
λ3	+	+	–	–	?	–	–	–	?
μ1	+	+	–	+	–	–	–	+	–
μ2	?	–	?	–	–	–	–	–	+
σ1	–	+	–	–	–	+	–	?	–
σ2	+	–	–	–	–	–	?	–	+
σ3	–	+	–	+	–	?	–	+	+

+, well-supported; ?, suspected; –, no current evidence

In analogy with rotaviruses (ESTES 1996), it is possible to describe the reovirus virion as having three concentric protein layers (Fig. 2), in each of which the proteins are arranged with icosahedral symmetry: an outer layer formed by σ3 (rotavirus analogue VP7), a middle layer formed by λ2 and μ1 (rotavirus analogue VP6), and an inner layer formed by λ1 and σ2 (rotavirus analogue VP2). At least two additional viral proteins project from these layers in either outward or inward directions: the cell attachment protein σ1 (rotavirus analogue VP4) projects out from the particle surface by more than 400 Å (FURLONG et al. 1988) and the RNA

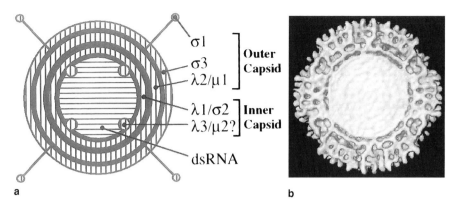

a   b

Fig. 2. a Cross-section of the reovirus virion as if sliced through an icosahedral twofold axis. Three concentric protein layers (*vertically hatched*) – σ3, λ2/μ1, and λ1/σ2 – are shown surrounding the dsRNA genome (*horizontally hatched*). Regions of lower density between the main protein layers are also shown (*solid gray*). The outwardly projecting σ1 protein is shown in its extended form, but not to scale for its length (too short). The structure that protrudes inward beneath the λ1/σ2 layer is thought to contain proteins λ3, μ2, and an N-terminal portion of λ1 (Dryden et al., submitted; see also Fig. 4). Reovirus virions are commonly described as having a two-layered structure including outer and inner capsids, as indicated. **b** Cross-sectional view of the reovirus T1L virion obtained by slicing the visualization of the cryoelectron density map (Dryden et al. 1993) through an icosahedral twofold axis

polymerase protein λ3 (rotavirus analogue VP1) projects into the particle interior by approximately 70 Å (Dryden et al., submitted). The remaining structural protein, μ2, may also be bound in the inwardly projecting structures (Dryden et al., submitted). Not counting the projections, the outer radius of reovirus virions is approximately 430 Å, and the total thickness of the protein layers is approximately 190 Å (Harvey et al. 1981; Metcalf et al. 1991; Dryden et al. 1993). The icosahedral symmetries of the protein layers are described in other sections (e.g., Sects. 1.4, 2.2, and 2.3).

One potential flaw with describing reovirus virions as having three main protein layers is that the different molecules of σ3 protein make only limited contacts over the particle surface and thus do not interlink to form a lattice (Figs. 3a, 4). It may thus be better to describe these particles as having only two main protein layers, namely, outer and inner capsids (Fig. 2). According to this view, both σ3 and σ1 are proteins that decorate the μ1–λ2 layer (Figs. 3a, b, 4). Another issue with describing σ3 as forming an independent layer in reovirus virions is that the σ3 and μ1 proteins show significant interdigitation across their lower and upper radii, respectively (Figs. 3a, b, 4, 5). Recognizing this fact, one might suggest that these interdigitating regions of σ3 and μ1 are most analogous to the rotavirus outer protein layer (VP7), whereas the lower regions of μ1 are most analogous to the rotavirus middle protein layer (VP6). Drawing the proper analogy may be important for allowing useful comparisons of the functions and/or assembly of these layers in the two groups of viruses. In any case, accepting the description of reoviruses as having only outer and inner capsids, the thicknesses of these layers, not including projections, are 125 and 65 Å, respectively (Dryden et al. 1993).

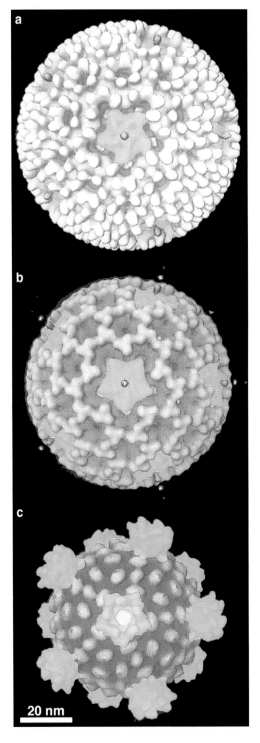

Fig. 3a–c. Surface-shaded views of **a** the reovirus T1L virion, **b** infectious subvirion particle (ISVP), and **c** core. Data was obtained by cryoelectron microscopy and three-dimensional image reconstruction (DRYDEN et al. 1993). Resolutions were determined to be 27, 28, and 32Å for the virion, ISVP, and core, respectively (DRYDEN et al. 1993). Each particle is viewed down a fivefold axis. Images were radially color-cued during visualization with the IRIS Explorer software package (Silicon Graphics) (SPENCER et al., 1997) and additional color coding was then added using Adobe Photoshop (Adobe Systems). As a result, the σ3 protein is shown in *white* (higher radii) to *light blue* (lower radii), the µ1 protein in *light blue* (higher radii) to *dark blue* (lower radii), the λ2 protein in *cyan*, the σ1 protein in *red–orange*, the λ1/σ2 proteins (core shell) in *lavender* (higher radii) to *magenta* (lower radii), and the putative λ3 protein in *yellow*

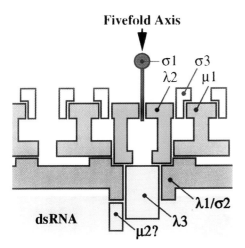

Fivefold Axis

σ1   σ3
λ2   μ1

λ1/σ2

dsRNA

μ2?   λ3

Fig. 4. Cross-sectional diagram of one fivefold region in the reovirus virion. Proteins are colored to approximate the coding in Fig. 3. The σ1 protein is drawn in its extended form but not to scale for its length. In some cases, the individual blocks are intended to represent protein monomers (e.g., σ3), but in other cases they represent homo-oligomers (e.g., μ1) or hetero-oligomers (λ1/σ2). Contacts across the base of the μ1 layer are not demonstrated here. The locations of the λ3 and μ2 proteins, and the internally directed region of λ1, are those suggested by DRYDEN et al. (submitted)

Special discussion is needed for the μ1 protein. It appears that this protein undergoes autoproteolytic cleavage after it assumes the necessary conformation during the assembly of virions (NIBERT et al. 1991b). The resulting fragments are named μ1N (amino-terminal fragment, molecular weight 4200) and μ1C (carboxy-terminal fragment, molecular weight 72,100), and both are found in virions in what are thought to be stoichiometric amounts (approximately 600 copies). A small amount of uncleaved μ1 protein is also found in virions, bringing the total of cleaved and uncleaved molecules to 600 copies; however, it is not currently considered that these uncleaved μ1 molecules have any special structural or functional significance. Both μ1 and μ1N are N-myristoylated at their amino termini (NIBERT et al. 1991b; TILLOTSON and SHATKIN 1992). For simplicity in this review, the term μ1 is generally used to refer to both the cleaved and uncleaved forms.

2.2 Specialized Regions Around the Fivefold Axes

As another way to simplify description of the reovirus particle, we can distinguish the regions of each concentric layer that surround the icosahedral fivefold axes versus the remaining regions, which are nearer the icosahedral threefold and twofold axes (Figs. 3, 4). In particular, the outwardly projecting σ1 protein and inwardly projecting λ3 (and possibly μ2) proteins are found only at the icosahedral fivefold axes. Similarly, the λ2 protein is found only at positions surrounding the fivefold axes, where it substitutes for molecules of μ1 in the T = 13 lattice (15 molecules of μ1 substituted by five molecules of λ2 at each fivefold axis). Thus the peripentonal regions of the reovirus capsids are "specialized" relative to the rest of the capsids. It is interesting to note that several critical functions for infections of cells are mediated by these fivefold-restricted proteins, including attachment to cell surface receptors (σ1; LEE et al. 1981), transcription (RNA polymerase, λ3; BRUENN 1991; STARNES and JOKLIK 1993), and mRNA capping (RNA guanylyltransferase, λ2; CLEVELAND et al. 1986; FAUSNAUGH and SHATKIN 1990; MAO and JOKLIK 1991).

LIVERPOOL JOHN MOORES UNIVERSITY
LEARNING SERVICES

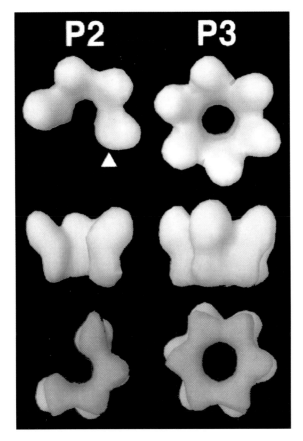

Fig. 5. P2 and P3 arrays of σ3 protein from reovirus T1L virions. The σ3 arrays were visualized in isolation by subtracting the cryoelectron density map of infectious subvirion particles (ISVP) from that of virions (DRYDEN et al. 1993), and single arrays were isolated by cropping the visualization of the difference map. The displayed arrays may contain some density not attributable to σ3. The arrays were smoothly rotated for display so that they appear as viewed from outside the virion at the top; from the side, within the σ3 layer, in the middle; and from inside the virion at bottom. Individual σ3 subunits – four in the P2 array and six in the P3 array – are evident from their characteristic morphological domains. The arrays were radially color-cued (SPENCER et al., 1997) so that the morphological domain that projects above the μ1 layer in virions is *white* and the one that projects into the μ1 layer in virions is *blue*. The open side of the P2 array is where λ2 substitutes the μ1 lattice in virions. The σ3 subunit that appears to contact λ2 is indicated with an adjacent *white triangle* in the top image of the P2 array

2.3 Radial Channels Through the Protein Layers

Proteins μ1 and λ2 are arranged in the reovirus outer capsid such that their masses are centered around local threefold axes in the T = 13 lattice (Fig. 3b). As a consequence of this and other aspects of the subunit arrangements, 132 large solvent-filled channels (25–110 Å across at different radii and in different particle types) are left to perforate the outer capsid at the 12 P1, 60 P2, and 60 P3 positions of the lattice (METCALF et al. 1991; Fig. 6). These channels help to explain the large

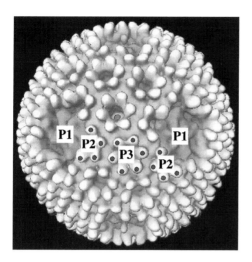

Fig. 6. Surface-shaded view of the reovirus T1L virion highlighting the P1–P3 solvent channels in the T = 13 (laevo) icosahedral lattice. Visualization from the cryoelectron density map (DRYDEN et al. 1993) was radially color-cued (SPENCER et al., 1997) to aid in identifying symmetrically equivalent features, but is shown here in *grayscale*. Particle is viewed down a twofold axis. The labeled channels trace out the path from one penta-coordinated site (*P1*), across three-hexa-coordinated sites (*P2* and *P3*), and through a left-hand turn to another penta-coordinated site (*P1*), the path characteristic of a T = 13(laevo) lattice. The four or six σ3 subunits surrounding the P2 or P3 channels, respectively, are labeled with *black dots*. The left-handed nature of the T = 13 lattice was imposed of the visualization because of other analyses (METCALF 1982; CENTONZE et al. 1995)

hydration state of this layer (HARVEY et al. 1974). Additional details relating to these channels are as follow. The σ3 protein interdigitates with μ1 around the perimeters of the P2 and P3 channels in virions, thus narrowing them at higher radii (Figs. 3a, b, 4). The presence of either four or six σ3 subunits surrounding the P2 or P3 channels, respectively, is illustrated especially well in Fig. 5. The 60 P3 channels appear open through the full thickness of the outer capsid, except that they are partially obscured by density near the base of μ1 in virions (DRYDEN et al. 1993; Fig. 6). The 60 P2 channels are partially obscured by regions of λ2 that project into them (Figs. 3a, b, 6). The P1 channel is open only at lower radii within the μ1/λ2 layer in virions and ISVPs (Fig. 7). It is closed at higher radii in virions and ISVPs because the outer regions of the five surrounding λ2 subunits closely approach the fivefold axis and because the remaining narrow channel is filled by the base of the σ1 fiber (DRYDEN et al. 1993; Figs. 3a, b, 4). In cores, however, the P1 channel is open through the λ2 protein (Fig. 3c; see Sect. 3.1) consequent to a conformational change involving the outer regions of λ2 and loss of the σ1 fiber (DRYDEN et al. 1993). It is interesting to note that, in rotaviruses, the P1 channel is open in virions because the cell attachment protein VP4, the equivalent of σ1, inserts into the P2 channels instead (YEAGER et al. 1990; SHAW et al. 1993).

The arrangement of proteins in the reovirus inner capsid remains more poorly defined, but in analogy with other dsRNA viruses (CHENG et al. 1994; SHAW et al. 1996), it seems likely that the λ1 protein (120 copies) is arranged as asymmetric

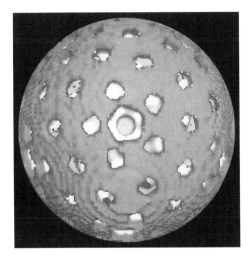

Fig. 7. Radial section of reovirus T1L virion through the base of the µl protein layer. A visualization from the cryoelectron density map (DRYDEN et al. 1993) was radially sectioned such that density at radii greater than 320 Å was removed from the displayed image. Visualization was radially color-cued (SPENCER et al., 1997) to aid in identifying symmetrically equivalent features, but is shown here in *grayscale*. The particle is displayed to approximate the scale with that in Fig. 3, but here the particle is viewed down a fivefold axis. The image shows µl forming an approximately solid layer at this radius except where it is perforated by the P1–P3 channels. Note that the P1 (fivefold) channel is wide open at this radius, which is below the radius at which the outer regions of λ2 (and the inserted σl protein) overhang it

dimers in a T = 1 structure. It also seems likely that this arrangement of subunits leaves a channel through the inner capsid at each fivefold axis, but that this channel is mostly filled by insertion of one or more proteins (λ3 and perhaps µ2) that form the inwardly projecting structure that is anchored to the inner capsid at each fivefold site (DRYDEN et al. submitted; Fig. 3c, 4). Small pores may also be present around the periphery of this fivefold channel and may serve as sites of exit for mRNA as they are newly transcribed by the viral RNA polymerase (SPENCER et al. 1997; YEAGER et al. 1996), similar to the mechanism recently described for rotaviruses (LAWTON et al. 1997).

2.4 Centrally Condensed dsRNA Segments

The ten genomic dsRNA segments of reoviruses (23,549 base pairs in total; WIENER et al. 1989) are packed into the center of the reovirus virion (radii, < 240 Å) and define another layer in its structure (Figs. 2, 4). Although contacts with the surrounding inner-capsid proteins are likely to occur near the periphery of this RNA sphere, particularly near the fivefold axes of the inner capsid (see Sect. 4.18; Figs. 2, 4), most of the RNA appears not to be protein associated. Instead, the individual dsRNA helices appear to be closely packed next to others in locally parallel arrays (26 Å separating the centers of adjacent helices; HARVEY et al. 1981; DRYDEN et al.

1993), most likely in a hexagonal arrangement (LEPAULT et al. 1987). This arrangement can also be described as a nematic liquid crystal (LEPAULT et al. 1987).

3 Samples and Methods for Structural Studies of Reovirus Particles

The sources of data for current conclusions about the structure of reovirus particles are described below. Although emphasis is placed on particle-oriented studies, studies of reovirus proteins outside the setting of virus particles, such as after expression in foreign systems, are contributing other important data. Good examples are studies by several investigators on the structure of σ1 (e.g., DERMODY et al. 1990; FRASER et al. 1990; STRONG et al. 1991).

3.1 Types of Reovirus Particles

To understand much of the literature on reovirus structure, one must appreciate the ready capacity of reovirus virions to undergo partial uncoating to yield subvirion particles. Especially significant are ISVPs and cores, the structures of which are characterized by the loss of outer-capsid proteins σ3 or σ3, σ1, and μ1, respectively, in addition to other changes (Table 3; Fig. 3). Commonly, ISVPs and cores are generated from virions by in vitro treatments with alkaline proteases, e.g., chymotrypsin; however, an important reason for interest in these particles, in ad-

Table 3. Status of outer-capsid proteins in reovirus virions, infectious subvirion particles (ISVPs), and cores[a]

Protein	Particle		
	Virion	ISVP	Core
λ2	Outer region parallel to the particle surface; points toward fivefold axis and contacts σ1; P1 channel closed	Outer region parallel to the particle surface; points toward fivefold axis and contacts σ1; P1 channel closed	Outer region rotated to point more perpendicular to the particle surface; P1 channel open
μ1	Most μ1 cleaved into μ1N and μ1C; small amount intact	Most μ1 cleaved into μ1δ and φ; most μ1C cleaved into δ and φ; μ1N present	Missing: degraded
σ1	May adopt folded form in some isolates	Adopts extended form that projects from the particle surface; cleaved within fiber region in some isolates	Missing: eluted
σ3	Intact	Missing: degraded	Missing: degraded

[a] Sources of information referenced in text.

dition to their use in studies of reovirus structure, is that related particle forms appear to be generated during reovirus entry into mammalian hosts (BODKIN et al. 1989) and/or single cells (e.g., STURZENBECKER et al. 1987) and to play essential roles in entry (reviewed by NIBERT et al. 1991a). Thus structural changes in ISVPs and cores are thought to reflect "programmed virus disassembly" and to point toward features of reovirus structure that are important for initiating infection.

When purifying reovirus virions from infected cells, one obtains not only infectious virions but also a substantial yield of top component virions, which have a nearly identical complement of proteins as do virions, but are noninfectious because they lack the viral genome (SMITH et al. 1969). Although the mechanisms by which they arise remain largely unknown, top component virions have also been useful for structural studies, as have top component ISVPs (LUCIA-JANDRIS et al. 1993) and top component cores (SMITH et al. 1969; WHITE and ZWEERINK 1976) derived from them by treatments with alkaline proteases analogous to those on virions. Top component particles were recently used to provide evidence localizing the reovirus RNA polymerase inside particles (DRYDEN et al., submitted).

In addition to the preceding six particle forms, a variety of treatments have been used to remove specific proteins or regions of proteins from those particles in an effort to learn more about reovirus structure. Examples include the use of sodium dodecyl sulfate (SDS) to remove σ3 from virions (DRAYNA and FIELDS 1982), heat to remove σ1 from ISVPs (FURLONG et al. 1988), and high pH to remove λ2 from cores (WHITE and ZWEERINK 1976). A related approach is to observe the structure of reovirus particles in the process of one of their functions, such as the recent study of cores transcribing the viral mRNA (YEAGER et al. 1996). Partially assembled particles obtained from infected cells (MORGAN and ZWEERINK 1975; ZWEERINK et al. 1976) may also be useful for studies of reovirus structure as well for expanding our currently sketchy knowledge of reovirus morphogenesis. Temperature-sensitive mutants for which assembly at nonpermissive temperatures is blocked at particular steps may be important sources of such particles (MORGAN and ZWEERINK 1974; HAZELTON and COOMBS 1995). Other recent findings suggest the utility of particles assembled from cDNA-expressed proteins for studies of reovirus structure and assembly (XU et al. 1993).

3.2 Biophysical and Other Techniques

Information about the structure of reovirus particles comes from a variety of sources; however, atomic-scale models, as from X-ray crystallography or nuclear magnetic resonance spectroscopy, are not yet available for any reovirus particles or proteins. Crystals of reovirus cores diffracting to resolutions below 8 Å have been reported (COOMBS et al. 1990) and offer hope for an atomic-scale model of those particles in the future (K.M. Reinisch and S.C. Harrison, unpublished).

Three-dimensional image reconstructions derived from transmission cryoelectron micrographs of particles embedded in vitreous ice represent the highest-resolution images of reovirus particles to date. The original such data were provided by

Metcalf and colleagues for virions of reoviruses T2J and T3D and cores of T3D (METCALF et al. 1991). More recent studies by Baker, Yeager, and colleagues have provided similar images at higher resolutions for virions, ISVPs, and cores of reovirus T1L (DRYDEN et al. 1993), top component virions, ISVPs, and cores of T1L (DRYDEN et al. submitted), transcribing cores of T1L (YEAGER et al. 1996), and cores of T3D (LUONGO et al. 1997). The more recent images have resolutions below 30 Å, at which it is possible to discern symmetry relationships among the protein subunits, to distinguish morphological domains in some subunits, and to predict in some cases which morphological domains belong to which individual subunits within the repeating structures of the viral capsids. Images approaching similar resolutions for the surfaces of single reovirus particles were recently obtained by scanning cryoelectron microscopy (CENTONZE et al. 1995) and confirm many of the features observed in the image reconstructions from transmission cryoelectron micrographs.

Before data from cryoelectron microscopy, useful images of reovirus particles were obtained by ambient-temperature electron microscopy employing either negative staining or shadowing with heavy metals. These earlier experiments were important for defining the general features of reovirus particles, including the concentric protein layers, centrally condensed dsRNA, distribution of proteins among the layers, organization of the outer capsid as a T = 13(laevo) lattice (METCALF 1982), turret-like structures formed by the λ2 protein around the fivefold axes in cores (WHITE and ZWEERINK 1976), and fibers formed by the σ1 protein projecting from the fivefold axes in ISVP and some virions (FURLONG et al. 1988). These more routine methods of electron microscopy remain useful for analyzing reovirus structure.

A variety of other techniques have contributed important data toward understanding the structure of reovirus particles. Notable examples are biophysical studies by Bellamy and colleagues, which provided measurements for the molecular weights of reovirus virions and cores (FARRELL et al. 1974), evidence for the highly hydrated (porous) nature of the outer, but not inner capsid (HARVEY et al. 1974), and evidence for the packing of the dsRNA genome in parallel arrays in the center of particles (HARVEY et al. 1981). Other good examples are studies with monoclonal antibodies to reovirus proteins, which provided evidence for the exposure of λ2 of the surface of virions (HAYES et al. 1981), different domains of σ1 (BURSTIN et al. 1982), and proximity of σ1 and σ3 on the surface of virions (VIRGIN et al. 1991). As mentioned in the preface to this section, various studies with expressed proteins are providing important information about reovirus structure, including data about protein–protein interactions such as evidence that an N-terminal region of σ1 anchors the protein to particles (LEONE et al. 1991) and evidence for pairwise interactions among all three λ proteins (STARNES and JOKLIK 1993).

4 Some Remaining Questions About the Structure of Reovirus Particles

Despite how much we seem to learn about reovirus particles, the number of questions do not decrease. Rather, as we gain more information and are able to think about more specific details of the structures, the questions increase in number. The following are a subset of current ones. While higher-resolution images obtained by cryoelectron microscopy or X-ray crystallography may answer many of these questions, some may require even more sophisticated approaches. Questions about how the structures are formed during particle morphogenesis are not discussed here, but little is known about those steps.

4.1 Contact Between the Base of the σ1 Fiber and λ2

Image reconstructions from cryoelectron micrographs of reovirus T1L virions and ISVPs suggest that the base of the σ1 fiber protrudes into the P1 channel such that it is contacted by the outermost regions of the five λ2 subunits that surround the channel (Dryden et al. 1993; Figs. 3a, b, 4). Since the fiber most likely represents a σ1 trimer (Strong et al. 1991; see Sect. 4.3), this arrangement implies a symmetry mismatch with the λ2 pentamer. As a result, there are questions concerning the nature of the σ1–λ2 contacts. For example, which specific residues mediate these contacts and from how many of the σ1 subunits and how many of the λ2 subunits? It seems relevant to consider that at the ISVP-to-core transition, when the outer regions of λ2 undergo a major change in conformation (Dryden et al. 1993; Fig. 3b, c), the σ1 protein is lost from particles. According to one interpretation, this finding suggests that contacts with a single λ2 subunit are not sufficient to keep σ1 bound to the particle; however, another possibility is that the binding sites for σ1 are destroyed or buried upon the conformational change in λ2. Also conceivable is an arrangement by which σ1 does not participate in any strong interactions with λ2, but instead remains tethered to virions and ISVP because a bulbous knob at the base of the σ1 fiber cannot escape through the top of the P1 channel until the channel is widened at the transition to cores.

4.2 Conformation of σ1 in Virions and Potential Contacts with σ3

Furlong et al. (1988) provided evidence from negative staining and electron microscopy that σ1 takes the form of an extended fiber projecting from the surface of some reovirus particles. In particular, projecting fibers were seen more often in virions of reovirus T2J than in those of T1L or T3D. Projecting fibers were also frequently seen in ISVPs, particularly those of T1L or T2J, suggesting that a change in σ1 conformation might occur at the virion-to-ISVP transition (D.B. Furlong and M.L. Nibert, unpublished; Furlong et al. 1988). Indeed, image reconstructions from cryoelectron micrographs reveal a difference between virions and ISVPs in the density attributed to σ1: the density extends further from the capsid in ISVPs, supporting the hypothesis that σ1 may be retracted in the virions of some isolates but extended in their ISVPs (DRYDEN et al. 1993; Fig. 3a, b). While these observations are intriguing and might have important implications for reovirus entry, more evidence is needed to confirm the retracted form of σ1. In the meantime, a fundamental question concerns what might be the structure of the retracted form. For example, might it involve major rearrangements of fiber sequences from those in the extended form (e.g., conversion between α-helix and β-sheet structures), bending at hinges within the fiber, or deeper protrusion of the fiber into the λ2 channel? Another question concerns whether the retracted form of σ1 might make additional contacts with outer-capsid proteins, particularly with σ3 (VIRGIN et al. 1991), to explain the proposed switch to the extended form in ISVPs.

4.3 Nature of the σ1 Oligomer

The strongest available data suggest that the reovirus T3D σ1 protein forms trimers (STRONG et al. 1991). Nevertheless, early cross-linking data suggested that T3D σ1 forms tetramers (BASSEL-DUBY et al. 1987). In addition, the twofold symmetry identified in image reconstructions from electron micrographs of negatively stained, isolated σ1 fibers from reovirus T2J ISVPs (FRASER et al. 1990) remains difficult to reconcile with the trimer model.

4.4 Regions of σ3 Structure and Binding Sites for μ1 and dsRNA

In image reconstructions of virions (DRYDEN et al. 1993), the σ3 protein appears to comprise two morphological domains: one that interdigitates with μ1 and one that projects above the μ1 layer (Figs. 2–5). In order to understand the structure and function of σ3, it will be useful to define the sequences that constitute these different morphological regions. In particular, it will be important to localize the σ3 sequences that contact μ1 in virions versus those that are involved in binding to dsRNA when σ3 is in solution (HUISMANS and JOKLIK 1976). This question holds interest because the former binding activity is thought to interfere with the latter and thereby to further regulate an effect of σ3 on translation involving the dsRNA-activated protein kinase (TILLOTSON and SHATKIN 1992). The current paradox is that amino-terminal sequences in σ3 appear to be required for binding to μ1 (DANIS et al. 1992; SHEPARD et al. 1996), whereas carboxy-terminal sequences appear to be required for binding to dsRNA (SCHIFF et al. 1988; MILLER and SAMUEL 1992; DENZLER and JACOBS 1994), suggesting that the regulation of these two activities is more complex than simple competition for overlapping binding sites (SHEPARD et al. 1996).

4.5 Cleavage Sites in σ3

Some data suggest that there is a defined cascade of protease cleavages leading to σ3 degradation at the virion-to-ISVP transition (MILLER and SAMUEL 1992; VIRGIN et al. 1994; SHEPARD et al. 1995; L.A. Schiff and M.L. Nibert, unpublished). This degradation is thought to be necessary before the underlying μ1 molecules can assist the reovirus particle in penetrating the membrane bilayer at the outset of infection (NIBERT and FIELDS 1992; LUCIA-JANDRIS et al. 1993; NIBERT 1993; TO-STESON et al. 1993; WESSNER and FIELDS 1993; HAZELTON and COOMBS 1995; HOOPER and FIELDS 1996). For understanding the virion-to-ISVP transition and regions of σ3 that must be lost prior to membrane penetration, it will be important to localize the different sites of σ3 cleavage within the protein and within virus particles. In addition, since interaction with μ1 permits σ3 to assume its protease-sensitive conformation (SHEPARD et al. 1995), it will be important to relate the cleavage sites to the regions of σ3 that interact with μ1 for defining how binding to μ1 might induce the protease-sensitive conformation of σ3.

4.6 Regions of μ1 Structure and Contacts Among μ1 Subunits

Especially in image reconstructions of ISVPs (DRYDEN et al. 1993), from which the obscuring density of σ3 has been removed, the interactions among μ1 subunits appear easily interpretable at higher radii: the subunits make intimate contacts within trimeric units (200 per particle) and more limited contacts between single subunits from adjacent trimers across sites of local twofold symmetry (Fig. 3b). At middle radii, the pattern is even simpler in that the intratrimer contacts are maintained but intertrimer contacts are absent, creating lateral pores through the μ1 layer at these radii (Fig. 2). At lower radii, near the core surface, however, the pattern is less interpretable in that individual subunits appear to participate in a larger number of intimate contacts with adjacent ones. As a result, the intratrimer interactions are difficult to distinguish from others and a highly cross-linked structure is formed. In order to understand the structure and function of μ1, it will be necessary to define the sequences of μ1 that constitute its different morphological regions (Fig. 4) and to define the pattern of subunit interactions in the lower third of the structure. As an example, in order to draw models about μ1-membrane interactions, it would be useful to know in which morphological region the N-myristoyl group of μ1 (NIBERT et al. 1991b) is located, as well as the μ1N–μ1C cleavage junction (JAYASURIYA et al. 1988; NIBERT et al. 1991b). Since mutations near either of these sites abrogate μ1 binding to σ3 in solution (TILLOTSON and SHATKIN 1992), the sites seem likely to be located in the upper third of μ1 structure, i.e., if the effects of these mutations are relatively direct.

4.7 The δ–φ Cleavage Site in μ1

In analogy with the cleaved fusion proteins of many enveloped viruses, the cleavage of protein μ1/μ1C to generate fragments μ1δ/δ and φ in ISVPs has seemed likely to be important for membrane penetration by reovirus particles (e.g., NIBERT and FIELDS 1992). It might thus be useful to pinpoint this cleavage site within virus particles as well. However, since recent data suggest that cleavage at the δ–φ junction may be dispensable for reovirus infection (CHANDRAN and NIBERT, 1998), this question assumes less immediate interest.

4.8 Structural Changes in μ1 Accompanying Membrane Penetration

A large rearrangement in the structure of μ1 accompanies interactions between reovirus particles and membrane bilayers in in vitro assays (K. CHANDRAN and M.L. NIBERT, unpublished; NIBERT 1993) and is thought to be required for membrane penetration during productive infection (NIBERT and FIELDS 1992; LUCIA-JANDRIS et al. 1993; TOSTESON et al. 1993; WESSNER and FIELDS 1993; HAZELTON and COOMBS 1995; HOOPER and FIELDS 1996). The structure of reovirus particles containing this penetration-competent conformer of μ1 remains unvisualized to date, but would seem to be essential for understanding this critical step in infection. The hydrophobic character of these particles (K. CHANDRAN and M.L. NIBERT, unpublished; NIBERT 1993) may represent a challenge to these structural characterizations.

4.9 Different Environments of Different μ1 and σ3 Subunits

Because the T = 13(laevo) lattice formed by μ1 is incomplete and substituted by λ2 around the fivefold axes, different μ1 subunits reside in three different environments within this layer (METCALF et al. 1991; DRYDEN et al. 1993). Two minor subsets (60 copies each) are those adjacent to λ2 in the lattice, and although both appear to contact λ2, they appear to do so via different regions in both μ1 and λ2 (Fig. 3b). Because σ3 binds to μ1 in the lattice, different σ3 subunits in virions also reside in three different environments. Again, two minor subsets (60 copies each) are those adjacent to λ2; however, in the case of σ3, only one of these appears to contact λ2 (Fig. 3a; Fig. 5, P2). The other has one side apparently open to solvent but possibly available to interact with the retracted form of σ1 suggested above. The existence of these different subsets of μ1 and σ3 is certain, but questions remain as to whether they might exhibit distinct properties.

4.10 Contacts Between μ1 Subunits and Core Nodules

SHAW et al. (1996) recently proposed a model for regular contacts between the T = 1 inner capsid and T = 13(laevo) outer capsid of aquareoviruses; however, a similar model has yet to be proposed for reoviruses. It will be important to define the nature of contacts between μ1 and the 150 nodules that project over the core surface (Fig. 3c) in order to address how these contacts contribute to maintaining interactions between the outer and inner capsids in virions and ISVPs (Fig. 4). This is especially interesting in light of some available evidence that the different core nodules have different compositions (DRYDEN et al. 1993; SPENCER et al., 1997).

4.11 Arrangement of λ1 and σ2 in the Inner Capsid

In contrast to findings for the outer-capsid subunits, it is difficult to define the boundaries between the subunits of inner-capsid proteins from the 30-Å maps obtained by cryoelectron microscopy and image reconstruction (DRYDEN et al. 1993). This is true even with cores (Fig. 3c). Thus the general arrangement and relative positions of proteins λ1 and σ2, which together form the inner-capsid lattice (WHITE and ZWEERINK 1976; XU et al. 1993), remain unknown. Recent evidence that several other dsRNA viruses contain 60 asymmetric dimers (120 copies) of one protein in the capsid that surrounds their genome (CHENG et al. 1994; SHAW et al. 1996) suggests that λ1 (120 copies) is an analogous protein. As a result, λ1 will likely be found to interact in asymmetric dimers that are arranged in a T = 1 structure to create the backbone of the reovirus inner capsid. An attractive possibility might then be that one or more of the three types of nodules that project over the core surface (150 total; Fig. 3c) is formed primarily by σ2; however, evidence that σ2 is found in a more internal position (WHITE and ZWEERINK 1976) and has a dsRNA-binding activity (SCHIFF et al. 1988; DERMODY et al. 1991) may argue against this possibility.

4.12 Transcriptase Complexes

Image reconstructions recently derived from cryoelectron micrographs of reovirus T1L top component particles reveal a pinwheel-like structure protruding beneath each fivefold axis of the inner capsid (DRYDEN et al., submitted; Figs. 3c, 4). Other data in that study identify proteins λ3 and μ2, and an N-terminal region of λ1, as probable components of these complexes (Fig. 4). The position of these structures and the identities of their probable protein components suggest that they represent the reovirus transcriptase complexes (BARTLETT et al. 1974; JOKLIK 1983; SHATKIN and KOZAK 1983; STARNES and JOKLIK 1993; YEAGER et al. 1996; YIN et al. 1996; NOBLE and NIBERT 1997a). Important remaining questions about these structures include the following:

– How is the polymerase protein λ3 oriented within this complex, e.g., where is the polymerase active site and where are the other regions of this large protein (BRUENN 1991; STARNES and JOKLIK 1993)?

- Where is µ2 located within this complex and with which proteins is it associated?
- Similarly, is an N-terminal "domain" of λ1, possessing sequence similarities to NTPases and RNA helicases (BARTLETT and JOKLIK 1988; NOBLE and NIBERT 1997a, b), indeed a member of this complex and, if so, where is it positioned?
- Finally (a question essential for understanding the reovirus transcription mechanism), is there indeed one dsRNA segment per transcriptase complex in genome-containing particles?

4.13 Structural Changes in the Inner Capsid Accompanying Transcriptase Activation

Transcriptase activation at the ISVP-to-core transition is thought to involve the elimination of a block to transcript elongation, before which only short, abortive transcripts can be produced by the transcriptase complexes inside reovirus particles (YAMAKAWA et al. 1982). According to a recent model, NTPase activities (putative helicase activities) exhibited by core proteins λ1 (NOBLE and NIBERT 1997a) and/or µ2 (NOBLE and NIBERT 1997b) may be required for elongation (or at least for promoter clearance), and the activation of these enzymes at the ISVP-to-core transition (BORSA et al. 1973; NOBLE and NIBERT 1997a) may be a direct cause of transcriptase activation. It will be important to determine whether there are discernible structural changes in the inner capsid associated with the activation of these enzymes (POWELL et al. 1984; YEAGER et al. 1996). In addition, it will be important to define how these enzymes are inhibited in virions and ISVPs, perhaps by interaction with µ1 in the outer capsid (NOBLE and NIBERT 1997a; Fig. 4). Other structural changes that might accompany transcriptase activation include the widening of pores for substrate entry to or mRNA exit from the core. It remains unknown whether the large rearrangement in the outer regions of λ2 that accompanies the ISVP-to-core transition (DRYDEN et al. 1993; Fig. 3b, c) is relevant to transcriptase activation.

4.14 Enzymatic Regions of λ2

The λ2 protein is known to mediate the guanylyltransferase activity (CLEVELAND et al. 1986; FAUSNAUGH and SHATKIN 1990; MAO and JOKLIK 1991) and is suspected to mediate one or both of the methyltransferase activities (SELIGER et al. 1987; KOONIN 1993) involved in mRNA capping by reovirus particles. A simple model would suggest that the catalytic sites of these enzymes are located near the base of each monomer in the λ2 turret, in a good position to add a cap to each mRNA as it is extruded by the RNA polymerase beneath each fivefold axis (see above; Fig. 4). It will be important to localize these enzyme sites in λ2 as well as the pathways for RNA transport between them and the polymerase in order to understand how the transcription and capping enzymes are juxtaposed in the core to permit the efficient production of capped mRNA.

4.15 Molecular Weight of Cores

FARRELL et al. (1974) reported the molecular weight of reovirus T3D cores as $(52.3 \pm 2.6) \times 10^6$ based on diffusion coefficients. A similar value was obtained by scanning transmission electron microscopy (D.B. FURLONG et al., unpublished). Assuming that one each of the ten genomic dsRNA segments is present in every core on average and that no small molecules make substantial contributions to the weight of the core, the weight of the genome can be subtracted from the weight of the core to yield the approximate weight of protein in the core. Since the conglomerate weight of the ten dsRNA segments of reovirus T3D is 15.1×10^6 according to their nucleotide sequences, the approximate weight of protein in the core appears to approximate 37.2×10^6. When the sequence-predicted molecular weights and estimated copy numbers of the five core proteins are used to calculate the weight of proteins in the core, however, a somewhat lower value is obtained: 33.5×10^6.

This analysis suggests that we have something yet to learn about the components of cores. The most likely explanations for the discrepancy between the measured weight of the core and that calculated from sequence-predicted molecular weights and estimated copy numbers of proteins and dsRNA segments include one or more of the following:

1. The molecular weight of $\lambda1$ is greater than 137,400, reflecting an error in the published sequence (BARTLETT and JOKLIK 1988; S. NOBLE et al., unpublished).
2. The $\mu2$ protein is present in more than 12 copies per core, reflecting an error in estimation of its copy number.
3. The $\sigma2$ protein is present in more than 120 copies per core, reflecting an error in estimation of its copy number.
4. Small components, e.g., cationic compounds associated with the dsRNA, make an important contribution to the weight of the core.
5. One or more of the dsRNA segments is present in more than one copy per core.

Note that if the missing mass is discovered in the first four possibilities, the data will indicate that the weight of RNA in reovirus cores approximates that provided by one complete set of the ten dsRNA segments.

4.16 Fundamental Structure of the Packaged dsRNA Segments

Although no data directly address the issue of the fundamental structure of the packaged dsRNA segments, the RNA genome segments within reovirus particles are thought to adopt an A-form double-stranded helix over most of their lengths, as is exhibited by dsRNA in solution at low salt concentrations (ARNOTT et al. 1972). In addition, although the dsRNA helices appear to be mostly packed in locally parallel arrays inside particles (HARVEY et al. 1981; DRYDEN et al. 1993), little is known about the specific details of this arrangement, such as which elements may be important for permitting adjacent helices to slide past each other while more distant regions of those segments are being copied into mRNA by the transcriptase complexes in the core (YAMAKAWA et al. 1982). Similarly, nothing is known in any direct fashion about the arrangement of individual dsRNA segments within the core, such as whether any RNA may be restricted to certain regions of the particle cavity or whether any RNA may have greater tendencies to interact with certain others within the hexagonal array.

4.17 Linkage of dsRNA Segments Within Reovirus Particles

Another issue has concerned whether or not the dsRNA segments might be joined end-to-end within reovirus particles. That the segments can be separated on gels or gradients after particle disruption argues against the idea of linkage; however, some electron micrographs have supported the idea by showing molecules, apparently RNA released from particles, that are circular in some cases and much larger than any single segment (GRANBOULAN and NIVELEAU 1967; KAVENOFF et al. 1975). It is conceivable that the segments might be linked at their ends (a) by base-pairing or other noncovalent RNA–RNA interactions that are lost upon disruption of particles, (b) by noncovalent interactions involving one or more cross-linking proteins, or (c) by phosphodiester bonds that are cleaved by a particle-bound endonuclease activity upon disruption. One attractive feature of end-to-end linkage of the dsRNA segments is that it would provide a ready mechanism for bringing segments into position for the reinitiation of transcription upon completion of each round (see Sect. 4.18). Discussion of other pros and cons is beyond the scope of this review; nonetheless, in the absence of any stronger evidence for linkage, it seems likely at present that the dsRNA segments are not linked end-to-end within particles.

4.18 Interactions Between dsRNA Segments and Core Proteins

Questions about RNA–protein interactions within the core hold clear importance for understanding the mechanism of transcription by reovirus particles; however, very little is yet known. With regard to transcription, the most significant interactions seem likely to occur in regions beneath the fivefold axes of the inner capsid (Figs. 2, 4). One or more other types of interaction may also occur between inner-capsid protein or proteins and RNA elsewhere around the periphery of the RNA sphere, but such interactions seem likely to be weak. Only $\lambda1$ and $\sigma2$, the primary components of the inner capsid, have been reported to bind dsRNA (SCHIFF et al. 1988; DERMODY et al. 1991; LEMAY and DANIS 1994), but recent evidence indicates that $\mu2$ can do so as well (B. SHERRY, personal communication).

In the region of the transcriptase complexes, i.e., near the fivefold axes according to current data, a series of complex interactions between RNA and protein, within a specific structural setting, seem likely to occur. In particular, the capacity of virions to generate abortive transcripts representing the 5′ ends of the legitimate mRNAs (YAMAKAWA et al. 1982) indicates that the promoter regions of the dsRNA segments are already positioned within the polymerase active site in virions, i.e., most likely bound to protein λ3. In fact, a short region of dsRNA in this region may be unwound. The immediately downstream region of each dsRNA template seems likely to be specifically positioned as well, i.e., in a good orientation to be fed to the polymerase once the elongation phase of transcription can proceed. According to recent data, this interaction may be mediated by λ1 (LEMAY and DANIS 1994; NOBLE and NIBERT 1997a) and/or μ2 (COOMBS 1996; YIN et al. 1996; NOBLE and NIBERT 1997b) and may relate to an enzymatic function (NTPase, putative helicase) of one or both those proteins (NOBLE and NIBERT 1997a, b). It is conceivable that there is also some specific interaction for directing the rewound dsRNA away from the polymerase after it has served as a template. Other questions concern RNA–protein interactions that may be important for the efficient initiation of each new round of transcription from a dsRNA segment. Defining both the structural and mechanistic aspects of all phases of transcription within reovirus particles remains an interesting challenge.

4.19 Number of dsRNA Segments Per Particle

The interior of reovirus particles encloses a defined volume in which at least one each of the ten genomic dsRNA segments must be packaged if every particle is to be potentially infectious (SPENDLOVE et al. 1970). According to low-angle X-ray diffraction analyses (HARVEY et al. 1981), the radius of the RNA sphere approximates 240 Å and its volume thus approximates 5.8×10^7 Å^3. For comparison, one can calculate the volume that 23,549 base pairs of dsRNA (total size of the reovirus genome; WIENER et al. 1989) should occupy if it is A-form in structure (2.81 Å translation per base pair; ARNOTT et al. 1972) and packed in locally parallel, hexagonal arrays with an average center-to-center distance between dsRNA helices of 26 Å (HARVEY et al. 1981; DRYDEN et al. 1993). A correction must also be included for the volume not occupied by dsRNA in a hexagonal array (HARVEY et al. 1981). The final prediction of this calculation is that the reovirus genome should occupy a volume of approximately 4.7×10^7 Å^3 or a sphere of radius approximately 224 Å.

Which assumptions in the calculation might thus account for the discrepancy between the predicted and measured volumes of the reovirus genome? One possibility is that the dsRNA inside reovirus particles better approximates an A'-form helix, with a rise per base pair of 3.00 Å (ARNOTT et al. 1972), in which case the predicted volume would be increased to 5.0×10^7 Å^3. Another possibility is that the correction applied for open space in the hexagonal array is too small. For example, it seems reasonable that there might be space at the very center of the particle that is not occupied by RNA. Another possibility is that there are proteins that project from beneath the inner capsid, into the RNA sphere, and occupy a significant volume. In fact, we have evidence that transcriptase complexes that project inward at the fivefold axes occupy a volume of approximately 0.2×10^7 Å^3 (DRYDEN et al., submitted). A final possibility, of course, is that there are more than ten genome segments per particle. Assuming that the dsRNA adopts an A-form helix, accepting the original correction for hexagonal packing, and correcting for the volume oc-cupied by the transcriptase complexes, the measured volume of the particle interior can account for approximately 28,200 base pairs of dsRNA or enough for ap-proximately two more average-sized segments per particle. Although the present author considers it unlikely that 12 segments are packaged per particle, the pos-sibility must clearly be acknowledged according to these calculations. The question of how many segments are present per particle has important implications for the mechanism of RNA packaging by reoviruses. It also determines whether there might be two transcriptase complexes in reovirus particles that are not associated with a dsRNA segment.

4.20 Oligonucleotides

Reovirus virions contain two classes of single-stranded RNA oligonucleotides (BELLAMY et al. 1972). One class comprises poly(A) oligonucleotides that are two to 20 bases in length and are generally present in about 850 copies per virion. The other class comprises oligonucleotides that begin 5'GC(U)(A), represent abortive transcripts from the genomic dsRNA segments, are two to nine bases in length, and are generally present in about 2000 copies per virion. Both classes are thought to be formed by the viral RNA polymerase during the final stages of particle morpho-genesis; however, how the polymerase can form two distinct classes of oligonu-cleotides remains unknown. One possibility might be that the two polymerase complexes that lack an associated dsRNA segment (see above) generate poly(A), whereas the other ten complexes generate abortive transcripts.

Other remaining questions concern where the oligonucleotides are located and how they are retained in reovirus particles. From available observations it seems likely that they remain present in ISVPs, but not in cores (JOKLIK 1972; FARRELL et al. 1974). Since their conglomerate molecular weight is as high as 5×10^6, or one third of the weight of the genome, the oligonucleotides should contribute a sub-stantial amount to the electron-density maps of virions and ISVPs (DRYDEN et al. 1993), even if they are not arranged with icosahedral symmetry. Given the available data, it seems most probable that they are found along with the genome in the interior of virions and ISVPs, but are released upon the generation of cores, pos-sibly reflecting the widening of pores in the inner capsid (see above). If this is the case, then the oligonucleotides represent additional components that must be ac-commodated within the limited volume of the particle interior, further decreasing the likelihood that more than ten dsRNA segments are packaged there and in-creasing the likelihood that there is a specific mechanism to ensure packaging of only one each of the ten segments within each assembling particle.

Acknowledgements. Many thanks to former members of the Fields laboratory for all their ideas and encouragement over the years. Similar thanks to current and past members of my own laboratory. Special acknowledgment goes to Bernie Fields for his enduring inspiration. Other special thanks go to Timothy S. Baker, Kelly A. Dryden, Deirdre B. Furlong, Stephen C. Harrison, Alasdair C. Steven, and Mark Yeager for insights into virus structures and structural methods. Thank you to Timothy S. Baker and colleagues for making available the cryoelectron density maps for reovirus particles and to Stephan M. Spencer and Jean-Yves Sgro for assistance with graphics. Recent work in my laboratory was supported by NIH grant R29 AI39533, DARPA contract 144 GE53 A34 7300 4, a grant from the Lucille P. Markey Charitable Trust to the Institute for Molecular Virology, University of Wisconsin-Madison, a Shaw Scientists Award from the Milwaukee Foundation, and a Steenbock Career Development Award from the Department of Biochemistry, University of Wisconsin-Madison. Additional assistance was provided by The Graduate School and College of Agricultural and Life Sciences, University of Wisconsin-Madison.

References

Ahne W, Thomsen I, Winton J (1987) Isolation of a reovirus from the snake, Python regius. Arch Virol 94:135–139

Arnott S, Hukins DW, Dover SD (1972) Optimised parameters for RNA double-helices. Biochem Biophys Res Commun 48:1392–1399

Bartlett JA, Joklik WK (1988) The sequence of the reovirus serotype 3 L3 genome segment which encodes the major core protein $\lambda1$. Virology 167:31–37

Bartlett NM, Gillies SC, Bullivant S, Bellamy AR (1974) Electron microscopy study of reovirus reaction cores. J Virol 14:315–326

Bassel-Duby R, Nibert ML, Homcy CJ, Fields BN, Sawutz DG (1987) Evidence that the $\sigma1$ protein of reovirus serotype 3 is a multimer. J Virol 61:1834–1841

Bellamy AR, Nichols JL, Joklik WK (1972) Nucleotide sequences of reovirus oligonucleotides: evidence for abortive RNA synthesis during virus maturation. Nature (Lond) 238:49–51

Bodkin DK, Nibert ML, Fields BN (1989) Proteolytic digestion of reovirus in the intestinal lumens of neonatal mice. J Virol 63:4676–4681

Borsa J, Copps TP, Sargent MD, Long DG, Chapman JD (1973) New intermediate subviral particles in the in vitro uncoating of reovirus virions by chymotrypsin. J Virol 11:552–564

Bottcher B, Kiselev NA, Stel'Mashchuk VY, Perevozchikova NA, Borisov AV, Crowther RA (1997) Three-dimensional structure of infectious bursal disease virus determined by electron cryomicroscopy. J Virol 71:325–330

Bruenn JA (1991) Relationships among the positive strand and double-stranded RNA viruses as viewed through their RNA-dependent RNA polymerases. Nucleic Acids Res 19:217–226

Burstin SJ, Spriggs DR, Fields BN (1982) Evidence for functional domains on the reovirus type 3 hemagglutinin. Virology 117:146–155

Centonze VE, Ya C, Severson TF, Borisy GG, Nibert ML (1995) Visualization of individual reovirus particles by low-temperature, high-resolution scanning electron microscopy. J Struct Biol 115:215–225

Chandran K, Nibert ML (1998) Protease cleavage of reovirus capsid protein $\mu1/\mu1C$ is blocked by alkyl sulfate detergents, fielding a new type of infectious subvirion particle. J Virol 72:467–475

Chappell JD, Goral MI, Rodgers SE, dePamphilis CW, Dermody TS (1994) Sequence diversity within the reovirus S2 gene: reovirus genes reassort in nature, and their termini are predicted to form a panhandle motif. J Virol 68:750–756

Cheng RH, Caston JR, Wang GJ, Gu F, Smith TJ, Baker TS, Bozarth RF, Trus BL, Cheng N, Wickner RB et al. (1994) Fungal virus capsids, cytoplasmic compartments for the replication of double-stranded RNA, formed as icosahedral shells of asymmetric Gag dimers. J Mol Biol 244:255–258

Cleveland DR, Zarbl H, Millward S (1986) Reovirus guanylyltransferase is L2 gene product $\lambda2$. J Virol 60:307–311

Coombs KM (1996) Identification and characterization of a double-stranded RNA-reovirus temperature-sensitive mutant defective in minor core protein $\mu2$. J. Virol 70:4237–4245

Coombs KM, Fields BN, Harrison SC (1990) Crystallization of the reovirus type 3 Dearing core. Crystal packing is determined by the $\lambda2$ protein. J Mol Biol 215:1–5

Danis C, Garzon S, Lemay G (1992) Further characterization of the ts453 mutant of mammalian orthoreovirus serotype 3 and nucleotide sequence of the mutated S4 gene. Virology 190:494–498

Denzler KL, Jacobs BL (1994) Site-directed mutagenic analysis of reovirus $\sigma3$ protein binding to dsRNA. Virology 204:190–199

Dermody TS, Nibert ML, Bassel-Duby R, Fields BN (1990) A σ1 region important for hemagglutination by serotype 3 reovirus strains. J Virol 64:5173–5176

Dermody TS, Schiff LA, Nibert ML, Coombs KM, Fields BN (1991) The S2 gene nucleotide sequences of prototype strains of the three reovirus serotypes: characterization of reovirus core protein σ2. J Virol 65:5721–5731

Drayna D, Fields BN (1982) Biochemical studies on the mechanism of chemical and physical inactivation of reovirus. J Gen Virol 63:161–170

Dryden KA, Wang G, Yeager M, Nibert ML, Coombs KM, Furlong DB, Fields BN, Baker TS (1993) Early steps in reovirus infection are associated with dramatic changes in supramolecular structure and protein conformation: analysis of virions and subviral particles by cryoelectron microscopy and image reconstruction. J Cell Biol 122:1023–1041

Duncan R (1996) The low pH-dependent entry of avian reovirus is accompanied by two specific cleavages of the major outer capsid protein μ2C. Virology 219:179–189

Duncan R, Horne D, Cashdollar LW, Joklik WK, Lee PWK (1990) Identification of conserved domains in the cell attachment proteins of the three serotypes of reovirus. Virology 174:399–409

Duncan R, Murphy FA, Mirkovic RR (1995) Characterization of a novel syncytium-inducing baboon reovirus. Virology 212:752–756

Estes MK (1996) Rotaviruses and their replication. In: Fields BN, Knipe DM, Howley PM (eds) Fields virology, 3rd edn. Lippincott-Raven, Philadelphia, pp 1625–1656

Farrell JA, Harvey JD, Bellamy AR (1974) Biophysical studies of reovirus type 3. I. The molecular weight of reovirus and reovirus cores. Virology 62:145–153

Fausnaugh J, Shatkin AJ (1990) Active site localization in a viral mRNA capping enzyme. J Biol Chem 265:7669–7672

Fraser RD, Furlong DB, Trus BL, Nibert ML, Fields BN, Steven AC (1990) Molecular structure of the cell-attachment protein of reovirus: correlation of computer-processed electron micrographs with sequence-based predictions. J Virol 64:2990–3000

Furlong DB, Nibert ML, Fields BN (1988) Sigma 1 protein of mammalian reoviruses extends from the surfaces of viral particles. J Virol 62:246–256

Goral MI, Mochow-Grundy M, Dermody TS (1996) Sequence diversity within the reovirus S3 gene: reoviruses evolve independently of host species, geographic locale, and date of isolation. Virology 216:265–271

Granboulan N, Niveleau A (1967) Etude au microscope electronique du RNA de reovirus. J Microsc 6:23–30

Harvey JD, Farrell JA, Bellamy AR (1974) Biophysical studies of reovirus type 3. II. Properties of the hydrated particle. Virology 62:154–160

Harvey JD, Bellamy AR, Earnshaw WC, Schutt C (1981) Biophysical studies of reovirus type 3. IV. Low-angle x-ray diffraction studies. Virology 112:240–249

Hayes EC, Lee PWK, Miller SE, Joklik WK (1981) The interaction of a series of hybridoma IgGs with reovirus particles. Demonstration that the core protein λ2 is exposed on the particle surface. Virology 108:147–155

Hazelton PR, Coombs KM (1995) The reovirus mutant tsA279 has temperature-sensitive lesions in the M2 and L2 genes: the M2 gene is associated with decreased viral protein production and blockade in transmembrane transport. Virology 207:46–58

Hewat EA, Booth TF, Roy P (1994) Structure of correctly self-assembled bluetongue virus-like particles. J Struct Biol 112:183–191

Hooper JW, Fields BN (1996) Role of the μ1 protein in reovirus stability and capacity to cause chromium release from host cells. J Virol 70:459–467

Huismans H, Joklik WK (1976) Reovirus-coded polypeptides in infected cells: isolation of two native monomeric polypeptides with affinity for single-stranded and double-stranded RNA, respectively. Virology 70:411–424

Jayasuriya AK, Nibert ML, Fields BN (1988) Complete nucleotide sequence of the M2 gene segment of reovirus type 3 Dearing and analysis of its protein product μ1. Virology 163:591–602

Joklik WK (1972) Studies on the effect of chymotrypsin on reovirions. Virology 49:700–801

Joklik WK (1983) The reovirus particle. In: Joklik WK (ed) The reoviridae. Plenum, New York, pp 9–78

Kavenoff R, Talcove D, Mudd JA (1975) Genome-sized RNA from reovirus particles. Proc Natl Acad Sci USA 72:4317–4321

Kedl R, Schmechel S, Schiff L (1995) Comparative sequence analysis of the reovirus S4 genes from 13 serotype 1 and serotype 3 field isolates. J Virol 69:552–559

Koonin EV (1992) Evolution of double-stranded RNA viruses: a case for polyphyletic origin from different groups of positive-stranded RNA viruses. Semin Virol 3:327–340

Koonin EV (1993) Computer-assisted identification of a putative methyltransferase domain in NS5 protein of flaviviruses and λ2 protein of reovirus. J Gen Virol 74:733–740

Lawton JA, Estes MK, Prasad BV (1997) Three-dimensional visualization of mRNA release from actively transcribing rotavirus particles. Nature Struct Biol 4:118–121

Lee PWK, Hayes EC, Joklik WK (1981) Protein σ1 is the reovirus cell attachment protein. Virology 108:156–163

Lemay G, Danis C (1994) Reovirus λ1 protein: affinity for double-stranded nucleic acids by a small amino-terminal region of the protein independent from the zinc finger motif. J Gen Virol 75:3261–3266

Leone G, Mah DC, Lee PWK (1991) The incorporation of reovirus cell attachment protein σ1 into virions requires the N-terminal hydrophobic tail and the adjacent heptad repeat region. Virology 182:346–350

Lepault J, Dubochet J, Baschong W, Kellenberger E (1987) Organization of double-stranded DNA in bacteriophages: a study by cryo-electron microscopy of vitrified samples. EMBO J 6:1507–1512

Liu HJ, Giambrone JJ (1997) Amplification, cloning and sequencing of the σC-encoded gene of avian reovirus. J Virol Methods 63:203–208

Lucia-Jandris P, Hooper JW, Fields BN (1993) Reovirus M2 gene is associated with chromium release from mouse L cells. J Virol 67:5339–5345

Luongo CL, Dryden KA, Farsetta DL, Margraf RL, Severson TF, Olson NH, Fields BN, Baker TS, Nibert ML (1997) Localization of a C-terminal region of λ2 protein in reovirus cores. J Virol 71:8035–8040

Mao ZX, Joklik WK (1991) Isolation and enzymatic characterization of protein λ2, the reovirus guanylyltransferase. Virology 185:377–386

Metcalf P (1982) The symmetry of the reovirus outer shell. J Ultrastruct Res 78:292–301

Metcalf P, Cyrklaff M, Adrian M (1991) The three-dimensional structure of reovirus obtained by cryo-electron microscopy. EMBO J 10:3129–3136

Miller JE, Samuel CE (1992) Proteolytic cleavage of the reovirus σ3 protein results in enhanced double-stranded RNA-binding activity: identification of a repeated basic amino acid motif within the C-terminal binding region. J Virol 66:5347–5356

Morgan EM, Zweerink HJ (1974) Reovirus morphogenesis. Corelike particles in cells infected at 39° with wild-type reovirus and temperature-sensitive mutants of groups B and G. Virology 59:556–565

Morgan EM, Zweerink HJ (1975) Characterization of transcriptase and replicase particles isolated from reovirus-infected cells. Virology 68:455–466

Murphy FA (1996) Virus taxonomy. In: Fields BN, Knipe DM, Howley PM (eds) Fields virology, 3rd edn. Lippincott-Raven, Philadelphia, pp 15–57

Ni Y, Ramig RF, Kemp MC (1993) Identification of proteins encoded by avian reoviruses and evidence for post-translational modification. Virology 193:466–469

Nibert ML (1993) Structure and function of reovirus outer-capsid proteins as they relate to early steps in infection. PhD Thesis. Harvard University, Cambridge, MA

Nibert ML, Fields BN (1992) A carboxy-terminal fragment of protein μ1/μ1C is present in infectious subvirion particles of mammalian reoviruses and is proposed to have a role in penetration. J Virol 66:6408–6418

Nibert ML, Dermody TS, Fields BN (1990) Structure of the reovirus cell-attachment protein: a model for the domain organization of σ1. J Virol 64:2976–2989

Nibert ML, Furlong DB, Fields BN (1991a) Mechanisms of viral pathogenesis. Distinct forms of reoviruses and their roles during replication in cells and host. J Clin Invest 88:727–734

Nibert ML, Schiff LA, Fields BN (1991b) Mammalian reoviruses contain a myristoylated structural protein. J Virol 65:1960–1967

Nibert ML, Chappell JD, Dermody TS (1995) Infectious subvirion particles of reovirus type 3 Dearing exhibit a loss in infectivity and contain a cleaved σ1 protein. J Virol 69:5057–5067

Noble S, Nibert ML (1997a) Characterization of an ATPase activity in reovirus cores and its genetic association with core-shell protein λ1. J Virol 71:2182–2191

Noble S, Nibert ML (1997b) Core protein μ2 is a second determinant of NTPase activities in reovirus cores. J Virol 71:7728–7735

Ozel M, Gelderblom H (1985) Capsid symmetry of viruses of the proposed birnavirus group. Arch Virol 84:149–161

Powell KF, Harvey JD, Bellamy AR (1984) Reovirus RNA transcriptase: evidence for a conformational change during activation of the core particle. Virology 137:1–8

Schiff LA, Nibert ML, Co MS, Brown EG, Fields BN (1988) Distinct binding sites for zinc and double-stranded RNA in the reovirus outer capsid protein σ3. Mol Cell Biol 8:273–283

Seliger LS, Zheng K, Shatkin AJ (1987) Complete nucleotide sequence of reovirus L2 gene and deduced amino acid sequence of viral mRNA guanylyltransferase. J Biol Chem 262:16289–16293

Shatkin AJ, Kozak M (1983) Biochemical aspects of reovirus transcription and translation. In: Joklik WK (ed) The reoviridae. Plenum, New York, pp 79–106

Shaw AL, Rothnagel R, Chen D, Ramig RF, Chiu W, Prasad BVV (1993) Three-dimensional visualization of the rotavirus hemagglutinin structure. Cell 74:693–701

Shaw AL, Samal SK, Subramanian K, Prasad BVV (1996) The structure of aquareovirus shows how the different geometries of the two layers of the capsid are reconciled to provide symmetrical interactions and stabilization. Structure 4:957–967

Shepard DA, Ehnstrom JG, Schiff LA (1995) Association of reovirus outer capsid proteins σ3 and μ1 causes a conformational change that renders σ3 protease sensitive. J Virol 69:8180–8184

Shepard DA, Ehnstrom JG, Skinner PJ, Schiff LA (1996) Mutations in the zinc-binding motif of the reovirus capsid protein σ3 eliminate its ability to associate with capsid protein μ1. J Virol 70:2065–2068

Smith RE, Zweerink HJ, Joklik WK (1969) Polypeptide components of virions, top component and cores of reovirus type 3. Virology 39(4):791–810

Spencer SM, Sgro J-Y, Dryden KA, Baker TS, Nibert ML (1997) IRIS Explorer software for radial-depth cueing reovirus particles and other macromolecular structures determined by transmission cryoelectron microscopy and three-dimensional image reconstruction. J Struct Biol 120:11–21

Spendlove RS, McClain ME, Lennette EH (1970) Enhancement of reovirus infectivity by extracellular removal or alteration of the virus capsid by proteolytic enzymes. J Gen Virol 8:83–94

Starnes MC, Joklik WK (1993) Reovirus protein λ3 is a poly(C)-dependent poly(G) polymerase. Virology 193:356–366

Strong JE, Leone G, Duncan R, Sharma RK, Lee PWK (1991) Biochemical and biophysical characterization of the reovirus cell attachment protein σ1: evidence that it is a homotrimer. Virology 184:23–32

Sturzenbecker LJ, Nibert M, Furlong D, Fields BN (1987) Intracellular digestion of reovirus particles requires a low pH and is an essential step in the viral infectious cycle. J Virol 61:2351–2361

Tillotson L, Shatkin AJ (1992) Reovirus polypeptide σ3 and N-terminal myristoylation of polypeptide μ1 are required for site-specific cleavage to μ1C in transfected cells. J Virol 66:2180–2186

Tosteson MT, Nibert ML, Fields BN (1993) Ion channels induced in lipid bilayers by subvirion particles of the nonenveloped mammalian reoviruses. Proc Natl Acad Sci USA 90:10549–10552

Vieler E, Baumgartner W, Herbst W, Kohler G (1994) Characterization of a reovirus isolate from a rattle snake, Crotalus viridis, with neurological dysfunction. Arch Virol 138:341–344

Virgin HW IV, Mann MA, Fields BN, Tyler KL (1991) Monoclonal antibodies to reovirus reveal structure/function relationships between capsid proteins and genetics of susceptibility to antibody action. J Virol 65:6772–6781

Virgin HW IV, Mann MA, Tyler KL (1994) Protective antibodies inhibit reovirus internalization and uncoating by intracellular proteases. J Virol 68:6719–6729

Weiner HL, Fields BN (1977) Neutralization of reovirus: the gene responsible for the neutralization antigen. J Exp Med 146:1305–1310

Wessner DR, Fields BN (1993) Isolation and genetic characterization of ethanol-resistant reovirus mutants. J Virol 67:2442–2447

White CK, Zweerink HJ (1976) Studies on the structure of reovirus cores: selective removal of polypeptide λ2. Virology 70:171–180

Wiener JR, Joklik WK (1989) The sequences of the reovirus serotype 1, 2, and 3 L1 genome segments and analysis of the mode of divergence of the reovirus serotypes. Virology 169:194–203

Wiener JR, Bartlett JA, Joklik WK (1989) The sequences of reovirus serotype 3 genome segments M1 and M3 encoding the minor protein μ2 and the major nonstructural protein μNS, respectively. Virology 169:293–304

Wilcox GE, Compans RW (1982) Cell fusion induced by Nelson Bay Virus. Virology 123:312–322

Xu P, Miller SE, Joklik WK (1993) Generation of reovirus core-like particles in cells infected with hybrid vaccinia viruses that express genome segments L1, L2, L3, and S2. Virology 197:726–731

Yamakawa M, Furuichi Y, Shatkin AJ (1982) Reovirus transcriptase and capping enzymes are active in intact virions. Virology 118:157–168

Yeager M, Dryden KA, Olson NH, Greenberg HB, Baker TS (1990) Three-dimensional structure of rhesus rotavirus by cryoelectron microscopy and image reconstruction. J Cell Biol 110:2133–2144
Yeager M, Weiner S, Coombs KM (1996) Transcriptionally active reovirus core particles visualized by electron cryo-microscopy and image reconstruction. Biophys J 70:A116
Yin P, Cheang M, Coombs KM (1996) The M1 gene is associated with differences in the temperature optimum of the transcriptase activity in reovirus core particles. J Virol 70:1223–1227
Zweerink HJ, Morgan EM, Skyler JS (1976) Reovirus morphogenesis: characterization of subviral particles in infected cells. Virology 73:442–453

Enzymatic and Control Functions of Reovirus Structural Proteins

Z. Yue and A.J. Shatkin

1 Introduction

Viruses have evolved in many diverse ways in order to utilize efficiently their limited genetic information and successfully overcome the various defense mechanisms of different hosts. For the multisegmented dsRNA-containing reoviruses, this has

Center for Advanced Biotechnology and Medicine, 679 Hoes Lane, Piscataway, NJ 08854-5638, USA

The early demonstration of reovirus gene reassortment by Bernie Fields provided fundamental insights into how reoviruses evolve, and this chapter is dedicated to his memory.

Table 1. Enzymatic and control functions of reovirus structural proteins

Protein	Genome segment	Capsid location	Functions
λ1	L3	Inner	NTPase/helicase, RNA triphosphatase
λ2	L2	Outer, also in core	Guanylyltransferase, methyltransferase? mRNA extrusion channel
λ3	L1	Inner	RNA-dependent RNA polymerase
μ1/μ1C	M2	Outer	Cell membrane interaction, apoptosis?
μ2	M1	Inner	Transcription and replication?
σ1	S1	Outer	Cell attachment, hemagglutination, inhibition of host DNA replication, interactions with cytoskeleton, apoptosis?
σ2	S2	Inner	Transcription?
σ3	S4	Outer	Translation control, inhibition of host RNA and protein synthesis, assembly of genome?

apparently been accomplished at the level of the structural proteins rather than the mRNA. In reovirus infections, only full-length, unspliced viral mRNA are produced and, with one exception, each contains a single translational open reading frame. However, most, if not all, of the virion structural proteins also have enzymatic activities and/or other functions that are essential in the reovirus life cycle (Table 1). This structural protein pleiotropy is summarized in this chapter, which is based on the many studies that have correlated virion proteins with functional activities and assigned them to specific reovirus genome segments.

2 Entry of Reovirus into Host Cells

The entry phase of reovirus infection includes several distinct steps: attachment to the cell surface, endocytic uptake, internalization, and uncoating (NIBERT et al. 1996). Virion outer-shell structural proteins σ1, μ1/μ1C, and σ3 are all involved in these early events. For example, numerous genetic and biochemical studies have shown that the entry of reoviruses into host cells is mediated primarily by interactions of μ1/μ1C and σ1 with cell surface components, including the membrane and putative receptor or receptors. A key step in the uncoating process is the proteolytic removal of major outer protein σ3. This process is mediated in vivo by one or more preexisting endosomal proteases and can be mimicked in vitro by chymotrypsin digestion of purified virions. The resulting subviral particles (ISVP), unlike cores that lack the outer-shell proteins (CARTER et al. 1974), have a specific infectivity similar to intact virions (JOKLIK 1972; SHATKIN and LaFIANDRA 1972), indicating that σ3, in contrast to σ1, is not essential for the productive entry of virus particles into cells. Although early steps of reovirus infection are generally described as receptor-mediated endocytosis followed by association with endosomes and lysosomes, details of the specific protein–membrane and protein–protein interactions that occur between reoviruses and different types of host cells remain to be defined.

2.1 Cell Attachment Protein σ1

A large body of work has demonstrated that σ1, the hemagglutinin on the virus surface, mediates virus binding to host cell receptor(s) (LEE et al. 1981a; WEINER et al. 1977, 1980; YEUNG et al. 1987). Thus, although σ1 is a minor virion constituent, it is the main determinant of reovirus serotype, infectivity, and tissue tropism. It is associated in the outer shell with σ3 and the pentameric λ2 protein and assumes a lollipop-like, head-and-tail fiber projecting from the surface (DRYDEN et al. 1993; HAYES et al. 1981; LUFTIG et al. 1972; METCALF et al. 1991; NIBERT et al. 1990; VIRGIN et al. 1991). Structural dissection of σ1 defined a virion anchorage domain in the N-terminal portion (FRASER et al. 1990; FURLONG et al. 1988; LEONE et al. 1991a; MAH et al. 1990) and a receptor-binding, C-terminal region (BRUCK et al. 1986; DUNCAN et al. 1991; NAGATA et al. 1987; STRONG et al. 1991; TURNER et al. 1992; WILLIAMS et al. 1991). Based on differences in specific infectivity and hemagglutination of virions and ISVP derived from different viral strains, it has been suggested that σ1 of the serotype 3 Dearing strain (T3D) contains more than a single receptor-binding domain (NIBERT and FIELDS 1992). In agreement with this suggestion, the N-terminal tail of σ1 was recently shown also to be important for receptor binding (CHAPPELL et al. 1997).

Although the erythrocyte surface protein glycophorin A can act as a reovirus receptor, as measured by direct σ1 binding (CHOI 1994; PAUL and LEE 1987), receptors on other reovirus-sensitive animal cells have not been identified. Available evidence implies that different regions of σ1 may be involved in binding to erythrocytes as compared to mouse L cells (NAGATA et al. 1987; YEUNG et al. 1989). By testing T3D σ1 protein binding to L cells, which contain one or more populations of saturable receptors (ARMSTRONG et al. 1984; EPSTEIN et al. 1984; GENTSCH and HATFIELD 1984; SAWUTZ et al. 1987), sialic acid was shown to be required for minimal binding (PAUL et al. 1989); it was further shown that a sialoglycoprotein was involved (GENTSCH and HATFIELD 1984). A number of other cell surface proteins have been reported to bind σ1, including β-adrenergic receptor-like protein (Co et al. 1985a, b), a 54-kDa protein on endothelial cells that recognizes T3D and serotype 1 Lang strain (TIL) virus (VERDIN et al. 1989), epidermal growth factor (EGF) receptor (TANG et al. 1993), and a 47-kDa protein from intestinal epithelial cells that recognizes TIL virus (WEINER et al. 1988). In addition, a p65–p95 complex from T lymphocytes (SARAGOVI et al. 1995) and neural cells (REBAI et al. 1996) has been described as binding T3D σ1.

An interesting model proposed to explain σ1–receptor interactions (PAUL and LEE 1987) includes initiation of binding by association with sialic acid followed by one or more specific surface receptors, resulting in a signal generated from the C-terminal receptor-binding region to the N-terminal region and conformational changes in σ1 as well as other capsid proteins. These changes might be essential for subsequent events, including endocytosis and penetration of virions into the host cytoplasmic compartment.

2.2 Interactions of µ1/µ1C with the Cell Membrane During Virus Penetration

There is general agreement that major outer-capsid protein µ1/µ1C plays an important role in reovirus penetration of the host cell membrane during virus entry. In intact virions, µ1C, which is derived from µ1 by a specific proteolytic cleavage, is the majority (approximately 90%) M2 gene product, but the resulting N-terminal fragment of µ1 is also apparently virion associated (Jayasuriya et al. 1988; Lee et al. 1981b; Nibert et al. 1991). µ1C and σ3 are present in 1:1 complexes that maintain the integrity and stability of the protective virion outer shell (Dryden et al. 1993; Lee et al. 1981b). Disassembly or uncoating of virus is usually preceded by, and may require, removal of σ3 to expose a membrane-interacting domain on µ1/µ1C. This capacity for membrane association probably lies in the intrinsic structural features of µ1/µ1C. First, it contains a conserved N-terminal N-myristoylation sequence and is modified both in vitro and in vivo by addition of a myristoyl group to the glycine residue in position 2 (see Sect. 6; Nibert et al. 1991; Raju et al. 1994; Tillotson and Shatkin 1992). Addition of fatty acid side groups increases hydrophobicity, which presumably facilitates interaction with the cell membrane (Hruby and Franke 1993). Second, several distinctly hydrophobic stretches are present in the N-terminal portion of µ1, and a predicted pair of long amphipathic α-helices flank a chymotrypsin/trypsin cleavage site in the C-terminal region (Jayasuriya et al. 1988; Nibert and Fields 1992). µ1C and µ1 both are cleaved in the C-terminal region to generate the larger N-terminal δ (or µ1δ) fragment and a smaller φ fragment. This proteolytic processing, which in vivo occurs in late endosomes and lysosomes (Chang and Zweerink 1971; Silverstein et al. 1972; Silverstein and Dales 1968), can be mimicked by protease treatment of purified virions (Borsa et al. 1973; Joklik 1972; Shatkin and LaFiandra 1972). The resulting ISVP contain stoichiometric amounts of δ (or µ1δ) and φ (Nibert and Fields 1992). Consistent with an important role for this proteolytic event, early steps in infection by intact virions, but not by ISVP, can be inhibited by NH_4Cl (Bodkin et al. 1989; Canning and Fields 1983; Maratos-Flier et al. 1986; Sturzenbecker et al. 1987). Since treatment with this weak base increases the pH within endosomes and lysosomes, it seems likely that processing of virions to ISVP is dependent on the action of one or more acid proteases in these endocytic compartments. In agreement with this finding, reovirus entry into mouse L cells is dependent on the vacuolar proton-ATPase activity, which can acidify endosomes and activate peptidases (Martines et al. 1996). Direct interaction of µ1/µ1C proteolytic products with the cell membrane was documented by the following observations: ISVP containing µ1δ/δ and φ, but not intact virions, permeabilized the membrane of host cells and resulted in release of preloaded Cr^{51} (Hooper and Fields 1996b; Lucia-Jandris et al. 1993). This permeabilization effect of ISVP was mapped to the M2 gene (encoding μ1) and was inhibited by anti-μ1 monoclonal antibody (HOOPER and FIELDS 1996a; LUCIA-JANDRIS et al. 1993). It was also suggested that the central region of μ1/μ1C, previously shown to be essential for virus stability to ethanol treatment (WESSNER and FIELDS 1993), was responsible for membrane permeabilization (HOOPER and FIELDS 1996a, b).

There has been considerable speculation about the exact mechanism or mechanisms by which reovirus and other nonenveloped viruses penetrate lipid bilayers. ISVP, processed from virions in endosomes, are released into the cytoplasm, where replication proceeds. This release from the vesicular compartment may occur through an induced membrane-spanner pore, since ISVP, but not intact virions or viral cores, have been shown to induce the formation of anion-selective channels in lipid bilayers (TOSTESON et al. 1993). In agreement with a membrane-penetrating role for processed μ1/μ1C, reovirus bearing a dominant temperature-sensitive (*ts*) lesion in the M2 gene was apparently trapped in endosomes at the nonpermissive temperature, resulting in decreased replication (HAZELTON and COOMBS 1995). Whether other structural proteins are also involved in membrane-related events early in reovirus infection is not clear, but λ2, σ3, and probably λ1 all interact with μ1/μ1C in intact virions, possibly influencing virus uptake.

3 Transcription and Capping of mRNA

After processed ISVP have exited the protease- and nuclease-rich endosomal compartment, synthesis of full-length plus strands (mRNA) by the virion-associated transcriptase occurs on the ten genomic double-stranded (ds)RNA templates by a process that is entirely conservative (BANERJEE and SHATKIN 1970; BORSA and GRAHAM 1968; LEVIN et al. 1970; SKEHEL and JOKLIK 1969). The ten different transcripts (three large, three medium, and four small), like each of the corresponding genomic plus strands, are 5' capped and methylated, but not 3' polyadenylated. Viral cores produced from purified virions by chymotryptic removal of μ1/μ1C, σ3, and σ1 are stable and contain all the activities necessary to produce capped and methylated transcripts, consistent with the inner capsid proteins having both enzymatic and structural functions (FURUICHI et al. 1975a; SHATKIN 1974). As predicted previously, viral cores also apparently contain ATPase (NOBLE and NIBERT 1997), NTPase/helicase (BISAILLON et al. 1997), RNA triphosphatase (FURUICHI et al. 1976; BISAILLON et al. 1997) and possibly other enzyme activities in addition to RNA polymerase, guanylyltransferase, and methyltransferase(s). Three major proteins (λ1, λ2, and σ3) and two minor polypeptides (λ3 and μ2) comprise the core. Image reconstructions of cryoelectron micrographs of cores revealed a λ2 "spike" protruding from the surface at each of the λ2 icosahedral vertices (DRYDEN et al. 1993). These spikes appear to be hollow and may serve as channels for viral nascent mRNA extrusion (BARTLETT et al. 1974). λ2 in transcribing cores, like the λ1 protein (POWELL et al. 1984), was conformationally altered and differed from the corresponding structures in intact virions (DRYDEN et al. 1993). Although structural details remain to be further defined, cores clearly represent an integral, multi-enzyme complex that can produce genomic transcripts in vitro for many hours or even days (SHATKIN and SIPE 1968).

It has been difficult, if not impossible, to solubilize functional proteins from cores (SHATKIN and KOZAK 1983). However, very useful correlations of core proteins and enzymatic activities have been obtained from genetic analyses with reassortant viruses (NOBLE and NIBERT 1997). By using biochemical and cloning procedures, λ2 (encoded by L2) was identified as the guanylyltransferase that caps mRNA 5′ ends (CLEVELAND et al. 1986; MAO and JOKLIK 1991; SHATKIN et al. 1983). The L3 polypeptide product λ1, like λ2, was shown to be labeled by pyridoxal phosphate during transcription, providing early evidence of a transcriptase catalytic site (MORGAN and KINGSBURY 1980). However, more recently, λ3 has been shown directly to possess RNA polymerase activity (STARNES and JOKLIK 1993). A role in transcription has also been suggested for the M1-encoded minor core protein μ2 based on reassortant analyses (DRAYNA and FIELDS 1982; YIN et al. 1996). Other recent findings indicate that λ1 in cores contains ATPase (NOBLE and NIBERT 1997) and NTPase/helicase activities (BISAILLON et al. 1997). λ1 could also be the RNA triphosphatase involved in mRNA capping (BISAILLON et al. 1997).

3.1 Transcriptase (dsRNA-Dependent RNA Polymerase)

Activation of the reovirus-associated transcriptase apparently requires release of conformational constraints on intact virions, e.g., by proteolytic removal of the outer capsid (SHATKIN and SIPE 1968) or by heat shock (BORSA and GRAHAM 1968). Removal of μ1/μ1C is apparently necessary for mRNA synthesis (BORSA et al. 1974; DRAYNA and FIELDS 1982; JOKLIK 1972; SHATKIN and LAFIANDRA 1972), but evidence suggests that ISVP and virions can produce short "initiator" oligonucleotides corresponding to the first few 5′-terminal residues in mRNA (SHATKIN and LAFIANDRA 1972; YAMAKAWA et al. 1982). It should be noted that viral cores, in addition to full-length mRNA, also produce short "initiator" transcripts (YAMAKAWA et al. 1981). Thus transcription may be a two-step process consisting of reiterative initiation followed by less efficient elongation of the nascent oligonucleotides to full-length transcripts. Some type of template checking may occur between initiation and elongation, because cores containing psoralen-cross-linked dsRNA produced diminished amounts of full-length transcripts rather than a spectrum of prematurely terminated products (NAKASHIMA et al. 1979). Biophysical and biochemical studies have demonstrated that the transcriptase in cores is in stable complexes with dsRNA and other proteins and is constitutively active with a temperature optimum of 47°–52 °C (KAPULER 1970). The transcriptase in ISVP can be activated by treatment with K^+, dimethylsulfoxide, and other solvents, temperature shifts, caffeine, theophylline, 5-bromodeoxyuridine (BUDR) and thymidine (TdR), probably by inducing a capsid conformational change (SCHIFF and FIELDS 1990).

Several observations indicate that reovirus transcription is regulated. During early infection, the predominant species of transcripts are copied from genome segments L1, M3, S3, and S4 (NONOYAMA et al. 1974; SPANDIDOS et al. 1976). Transcription of all ten segments follows, and this expansion of template usage can be inhibited by cycloheximide treatment, suggesting a requirement for newly synthesized protein(s) (LAU et al. 1975; WATANABE et al. 1968). Similarly, ISVP isolated from cycloheximide-treated cells synthesized ten viral mRNA in vitro, but became restricted to transcribing segments L1, M3, S3, and S4 upon reinfection of cycloheximide-treated cells (SHATKIN and LAFIANDRA 1972). The final copy numbers of each mRNA produced by the virion-associated transcriptase are also regulated (BANERJEE and SHATKIN 1970; JOKLIK 1981). Results from in vitro pulse-labeling experiments indicated that the three mRNA classes are synthesized by cores at rates in the order s > m > l, consistent with the relative yields of transcripts in vivo (BANERJEE and SHATKIN 1970). The basis for predominant or selective transcription of some genome segments during the early stage of infection and the more frequent transcription of the small segments relative to the large remains to be defined.

Many investigators have tried to decipher the details of dsRNA-dependent mRNA synthesis by viral cores, and evidence has been obtained for involvement of all five core proteins. Proteins $\lambda1$ and $\lambda2$ in transcribing particles were both labeled with pyridoxal 5′-phosphate, a nucleotide analogue that blocked the polymerase by binding noncompetitively and also inhibited all the mRNA-modifying enzymes (MORGAN and KINGSBURY 1980). This inhibition may reflect an indirect effect of analogue attachment to one or more core proteins that are involved in transcription, but not necessarily integral components of the RNA polymerase (MORGAN and KINGSBURY 1981). The early proposal that $\lambda1$ and $\lambda2$ form a polymerase catalytic site is less attractive in light of more recent work that $\lambda1$ and $\lambda2$ have NTPase/helicase and mRNA guanylyltransferase activities, respectively.

Minor core component $\lambda3$ is so far the only reovirus structural protein that has been demonstrated to possess polymerase activity in vitro (STARNES and JOKLIK 1993). Purified $\lambda3$ synthesized poly(G) on poly(C) templates, but failed to transcribe reovirus dsRNA into mRNA, or mRNA into minus strands. Consistent with the possibility of polymerase activity in $\lambda3$, genetic reassortant studies showed that genome segment L1, which encodes $\lambda3$, controls the polymerase pH optimum in each of the three different reovirus serotypes (DRAYNA and FIELDS 1982). Primary sequence analysis revealed that $\lambda3$ contains motifs including GDD that are conserved in all viral polymerases (BRUENN 1991; MOROZOV 1989). Attempts to reconstitute dsRNA- or ssRNA-dependent transcriptase activity in vitro using different combinations of the three λ proteins were unsuccessful (STARNES and JOKLIK 1993), emphasizing the importance of the core complex. In addition to the λ proteins, minor core protein $\mu2$ (encoded by the M1 gene), is also involved in transcription (YIN et al. 1996). The previously noted unusual temperature profile (KAPULER 1970) and differences in the transcriptase temperature optimum of the three viral serotypes have been ascribed to the M1 gene by reassortant analyses (YIN et al. 1996). Furthermore, like $\lambda3$, $\mu2$ is present in cores in approximately 12 copies, and both may be located at or near each fivefold axis. However, no direct interaction between λ3 and μ2 has been described, and μ2 contains no obvious polymerase consensus sequences (YIN et al. 1996). Some evidence is available to suggest that core protein σ2, like its avian reovirus counterpart σ1, is a subunit of the virion RNA polymerase (MARTINEZ-COSTAS et al. 1995). In addition, σ2 has been reported to bind dsRNA and contains a sequence similar to the β-subunit of *Escherichia coli* RNA polymerase (DERMODY et al. 1991). Clearly, a reconstituted system for dsRNA-dependent RNA synthesis is needed to establish which structural protein or proteins catalyze the formation of reovirus mRNA.

The precise intracellular site of reovirus transcription has also not been defined, but transcriptionally active, core-like particles have been observed located completely in the cytoplasm, i.e., not compartmentalized in vesicles (BODKIN and FIELDS 1989; BORSA et al. 1981). However, the possibility remains that transcription can be mediated by cores present in vacuoles (SILVERSTEIN and DALES 1968; STURZENBECKER et al. 1987), with nascent transcripts extruded into the cytoplasm through λ2 conduits extending through the vacuolar membrane.

3.2 Capping Enzyme λ2

Eukaryotic mRNA capping is a general phenomenon discovered by studying viral transcription. mRNA produced in vitro by reovirus cores, like the plus strands in genomic dsRNA (CHOW and SHATKIN 1975; MIURA et al. 1974), were found to contain "blocked" phosphates at the 5′ ends which were characterized as m^{7}GpppX "caps" (FURUICHI et al. 1975a, b) and subsequently shown to be present in most eukaryotic cellular and viral mRNA (SHATKIN 1976). Capping of reovirus mRNA occurred on short nascent oligomers (FURUICHI et al. 1976), suggesting that it is cotranscriptionally mediated by a core complex in which the polymerase and capping enzymes are closely associated. However, transcription and capping of reovirus mRNA (FURUICHI and SHATKIN 1976), like cellular mRNA (ERNST et al. 1983) can be uncoupled, and viral cores can cap preformed RNAs (FURUICHI and SHATKIN 1977).

From biochemical studies of many different viral and cellular systems, it has become clear that most, but not all eukaryotic mRNAs are 5′-terminally modified by the same mechanism as used by reoviruses (SHATKIN 1976). The cap on reovirus RNA consists of 7-methylguanosine in 5′-5′ linkage through three phosphate groups to the initial 2′-*O*-methyl guanosine (m^{7}GpppG^m) (FURUICHI et al. 1976). The 2′-O-methylated G is present in all plus strands of the dsRNA genomic segments as well as the ten mRNA species made in vitro or in vivo. In some reovirus mRNA isolated from infected cells, the G^m-adjacent residue, cytosine, was also 2′-O-methylated, probably by a cellular methyltransferase (LANGBERG and MOSS 1981). The mechanism of capped mRNA synthesis by reovirus-associated enzymes includes the following steps:

$$pppG + pppC \xrightarrow{\text{RNA polymerase}} pppGpC + PP_i$$

$$pppGpC \xrightarrow{\text{Triphosphatase}} ppGpC + P_i$$

$$pppG + ppGpC \xrightarrow{\text{Guanylyltransferase}} GppppGpC + PP_i$$

$$GppppGpC + SAM \xrightarrow{\text{Methyltransferase 1}} m^7GppppGpC + SAH$$

$$m^7GppppGpC + SAM \xrightarrow{\text{Methyltransferase 2}} m^7GppppG^mpC + SAH$$

where SAM is S-adenosylmethionine, SAH is S-adenosylhomocysteine, P_i is inorganic phosphate, and PP_i is inorganic pyrophosphate.

Polymerization of GTP and CTP precursor nucleoside triphosphates is first catalyzed by the RNA polymerase on the dsRNA templates. RNA triphosphatase next converts the 5' ends to diphosphate acceptors, followed by the covalent attachment of GMP from GTP by the guanylyltransferase. The resulting GppppG structure is then modified at the N^7 and 2'-OH positions by methyltransferase(s) that have not yet been assigned to specific reovirus structural protein(s). λ2 was the only polypeptide specifically labeled in purified virions by ultraviolet (UV) irradiation in the presence of ^{35}S-radiolabeled 8-azido-S-adenosylmethionine (SELIGER et al. 1987), suggestive of methyltransferase activity. Although amino acid sequence alignment indicated a significant, but limited homologous domain shared by λ2 and other viral and cellular methyltransferases (KOONIN 1993), recombinant λ2 that contained active guanylyltransferase failed to methylate reovirus mRNA caps, perhaps due to the absence of necessary interactions with other core proteins (MAO and JOKLIK 1991).

The pentameric spike protein λ2, which is encoded by the L2 gene, was identified as the capping enzyme on the basis of covalent radiolabeling with α-^{32}P-GTP (SHATKIN et al. 1983) and subsequent transfer of the labeled GMP to transcripts to form caps (CLEVELAND et al. 1986; MAO and JOKLIK 1991). The capping reaction thus proceeds in two distinct steps: formation of an enzyme-GMP intermediate followed by nucleotide transfer to the 5'-diphosphate end of the acceptor oligomer. The nucleotide attachment site on λ2 was localized to lysine 226 by [α-^{32}P]-GTP labeling followed by analysis of proteolytic fragments using sequence-directed antibodies as probes (FAUSNAUGH and SHATKIN 1990). As in other viral and cellular guanylyltransferases, GMP linkage was via a phosphoamide bond. The capping enzymes of several DNA viruses (vaccinia, Shope fibroma, and African swine fever virus) and two different yeasts (*Saccharomyces cerevisiae* and *Schizosaccharomyces pombe*) contain the guanylylated lysine in a conserved motif KXDG (SHUMAN et al. 1994), but the λ2 catalytic site (including K226 in the sequence KPTN) shares little or no similarity to these other capping enzymes, despite their essentially identical enzymatic functions. The reovirus guanylyltransferase apparently represents another capping enzyme family which may also include other dsRNA viruses. It is of interest in this regard that the simian rotavirus capping enzyme, VP3, contains the sequence KPTG (LIU et al. 1992), although it is not known whether guanylylation occurs at this site. Recombinant, vaccinia-expressed λ2, purified to homogeneity, was also shown to be reversibly guanylylated in the presence of GTP and capable of transferring GMP to 5′-diphosphate-containing reovirus mRNA to form cap structures, but RNA triphosphatase was not detected in the monomeric λ2 (Mao and Joklik 1991). This activity has apparently not been definitively mapped to a specific viral protein, despite the fact that nucleoside triphosphatase activity was detected in reovirus cores more than two decades ago (Borsa et al. 1970; Kapuler et al. 1970).

3.3 NTPase/Helicase Activities of λ1

Like DNA transcription, dsRNA-templated mRNA synthesis requires an energy-dependent helicase to unwind the duplexes, and recently it was reported that core protein λ1 has NTPase/helicase activity (Bisaillon et al. 1997). λ1 is a major inner-capsid protein which interacts with other core components, including σ2, λ2, and λ3 (Starnes and Joklik 1993; White and Zweerink 1976; Xu et al. 1993). Its amino acid sequence includes a putative zinc finger motif and nucleotide-binding site (Bartlett and Joklik 1988), and Zn^{2+} binding has been confirmed by blotting and affinity assays (Schiff et al. 1988; Bisaillon et al. 1997). λ1 binds nucleic acids, apparently nonspecifically (Lemay and Danis 1994). Core-associated AT-Pase activity has been mapped to the L3 segment encoding λ1 (Noble and Nibert 1997), and recombinant λ1 purified from yeast was also shown to hydrolyze ATP and dATP, and other nucleoside triphosphates to a lesser extent (Bisaillon et al. 1997). Consistent with a dsRNA-unwinding activity, λ1 contains consensus motifs found in other RNA/DNA helicases (Bisaillon et al. 1997).

4 Replication of the dsRNA Genome

Reovirus replication requires synthesis of ten unique dsRNA segments and accurate, specific assembly of one of each of the segments into newly formed particles by a process referred to as assortment. Assortment apparently occurs early in virus replication and involves mRNA plus strands generated by the parental virus. After one or more proteins associate with ("package") the assorted mRNA, minus-strand synthesis follows immediately (Antczak and Joklik 1992). Thus the transcripts from parental particles likely play a dual role in the reovirus life cycle: as mRNA for viral protein synthesis and as plus-strand templates for assortment and subsequent minus-strand synthesis.

Little is known about the signals needed for assortment and initiation of minus-strand synthesis. Alignment of plus-strand sequences indicated that all ten transcripts share short conserved regions at both the 5′ and 3′ termini (Antczak et al. 1982). Genetic studies of segment M1 deletion mutants identified sequences at both ends that may be responsible for replication and assembly (Zou and Brown

1992). Although the proteins and enzymatic activities involved in the sequential steps of dsRNA replication remain largely undefined, several structural and non-structural proteins were found in association with mRNA in single-strand RNA-containing complexes (ssRCC) implicated in assortment (ANTCZAK and JOKLIK 1992). λ3, in addition to transcriptase activity, may also be a replicase for the synthesis of minus strands (STARNES and JOKLIK 1993), and a recent analysis of gene segment M1 suggested that µ2 also plays a role in the replicative conversion of plus strands to progeny dsRNA (COOMBS 1996).

4.1 Viral Proteins Involved in Assortment and Packaging of Plus Strands

Nonstructural protein σNS (encoded by S3) has been implicated previously in assortment and packaging based on its capacity to bind ssRNA (HUISMANS and JOKLIK 1976). The isolation of RNA–protein complexes using specific antibodies indicated that plus strands within ssRCC are associated with two nonstructural proteins, σNS and µNS (encoded by M3), and one structural protein, the S4-encoded σ3 (ANTCZAK and JOKLIK 1992). Major outer-capsid protein σ3 binds dsRNA and not ssRNA, but binding was not sequence dependent (DENZLER and JACOBS 1994; HUISMANS and JOKLIK 1976). It was speculated that σ3 may recognize intrastrand hairpin loops that resemble dsRNA, but direct evidence is lacking for a functional importance of σ3 (or other ssRCC-associated proteins) in assortment and packaging. An attractive model for assembly of the tripartite dsRNA genome in bacteriophage φ6 by specific packaging of single-stranded template RNA has recently been described (QIAO et al. 1997), and it will be interesting to test whether the model applies generally to segmented dsRNA viruses.

4.2 Replicase: Potential Roles of Core Proteins λ3 and µ2 in the Synthesis of Minus Strands

Reovirus replicase is apparently a complex of proteins essential for recognition of plus-strand templates and enzymatic synthesis of minus strands. "Replicase particles" isolated from infected cells are about 40 nm in diameter, with sedimentation coefficients ranging from 180 to 550S. Structural proteins found in these particles include the core components λ1, λ3, and a lesser amount of λ2 relative to virions, together with altered amounts of outer-capsid proteins µ1/µ1C, σ1, and σ3 (JOKLIK 1995). Replicase particles are resistant to limited proteolytic digestion and nonionic detergent treatment, but sensitive to RNase. Once minus-strand synthesis has been completed and end-to-end duplexes formed, the dsRNA in nascent particles rapidly becomes resistant to RNase digestion (ACS et al. 1971). RNA replication requires protein synthesis (KUDO and GRAHAM 1966), and it was suggested that dsRNA synthesis occurs entirely within nascent core precursors of progeny virions, in agreement with the absence of free minus strands or dsRNA in infected cells (GOMATOS 1967).

Recombinant λ3 produced poly(G) on poly(C) templates, suggesting that it may be both the reovirus replicase and the transcriptase (STARNES and JOKLIK 1993; see Sect. 3.1). Other proteins and factors that participate together with λ3 in transcription and/or replication may modulate λ3 to produce dsRNA- or ssRNA-dependent RNA polymerase activity (STARNES and JOKLIK 1993). A recent report indicated that minor core protein μ2 may also be involved in the synthesis of minus strand (COOMBS 1996), since a *ts* mutant virus bearing a lesion in the M1 gene produced normal amounts of ssRNA in infected cells (or in vitro) at the nonpermissive temperature, but no detectable dsRNA.

5 Translation of Reovirus mRNA

Reovirus plus strands produced by the virion-associated transcriptase function in the cytoplasm of infected cells (or in vitro) as templates for the synthesis of viral proteins. Each of the ten reovirus mRNA codes for one polypeptide, except the S1 mRNA, which contains two open reading frames and is translated to produce a structural protein (σ1) and σ1NS, which is not found in virions (BELLI and SAMUEL 1993; FAJARDO and SHATKIN 1990b). Newly produced reovirus proteins can usually be detected within 2 h after infection, and at late stages most of the newly synthesized proteins are viral in origin (ZWEERINK and JOKLIK 1970). The precise kinetics of viral protein synthesis depends on the multiplicity of infection (MOI), the host cell type, and the temperature and pH of the environment (FIELDS and EAGLE 1973). Unlike many other viruses, there is no characteristic early-to-late translational switch, although transcription of four reovirus mRNA may occur before others.

Early work on reovirus mRNA translation provided many general insights into the mechanisms of eukaryotic protein synthesis (SHATKIN and KOZAK 1983). Reovirus mRNA become ribosome associated soon after they are transcribed, and the ribosome-binding sites have been identified (KOZAK 1977). It was shown that 40S ribosomal subunits attach to mRNA at or near the 5′-capped end and "scan" in an ATP-dependent mechanism to the first AUG in most cases (KOZAK and SHATKIN 1978; KOZAK 1981a). The 60S subunit then joins, and the AUG initiator and approximately 15 additional residues on either side form a ribosome-protected site (KOZAK and SHATKIN 1977). The sequence context of the initiator codon also apparently contributes to translation efficiency (KOZAK 1981b). In agreement with this notion, the two sites corresponding to the initiation regions for the σ1 and σ1NS proteins are utilized in the bicistronic S1 mRNA with different efficiencies (BELLI and SAMUEL 1993; FAJARDO and SHATKIN 1990a; KOZAK 1982). In addition, an AUG in the m1 mRNA which is not normally used for initiation can be artificially "activated" by altering its sequence context (RONER et al. 1993). Sequence context may also help explain why some reovirus mRNA are translated more frequently than others (SHATKIN and KOZAK 1983). The facilitating effect of the mRNA cap on ribosome binding and translation initiation, which was first observed with reovirus mRNA, has been observed with many other eukaryotic mRNA (KOZAK 1991; SHATKIN 1976).

Reovirus infection usually results in shutoff of host protein synthesis and the production of predominantly viral proteins (SHARPE and FIELDS 1982; ZWEERINK and JOKLIK 1970). The mechanism by which reovirus takes control of the cellular translation machinery is not well understood. At least two hypotheses have been proposed: (1) accumulating viral mRNA outcompete cellular mRNA for limiting translational component(s) or (2) a virus-encoded product actively switches translation from host to virus protein synthesis. Genetic and biochemical studies have implicated outer-capsid protein σ3 in the control of translation. For example, it was reported that σ3 can alter initiation and promote the translation of uncapped reovirus transcripts (SKUP and MILLWARD 1980a, b; SKUP et al. 1981). More recent studies indicate (as detailed in Sect. 7) that σ3 can influence translation by binding dsRNA and blocking activation of the dsRNA-dependent protein kinase (PKR).

6 Processing and Modification of Structural Proteins

Some reovirus proteins undergo conformational changes by taking advantage of cellular protein folding and modification systems to ensure active structures. These alterations have been shown to be necessary for specific functions during virus replication. For example, the cell attachment protein σ1, which usually forms homotrimers, contains independent oligomerization domains located in the N-terminal and C-terminal portions of the protein. These two regions were shown to be involved in two distinct mechanisms of trimerization (GILMORE et al. 1996; LEONE et al. 1996). In addition, major outer-capsid protein μ1 is N-myristoylated, probably cotranslationally, and also proteolytically cleaved to μ1C (NIBERT et al. 1991; TILLOTSON and SHATKIN 1992). These events are probably required for virus penetration early in infection and possibly for progeny virus assembly at later times.

6.1 Trimerization and Folding of σ1

The cell attachment protein and hemagglutinin σ1 binds to cell receptors and mediates the endocytic engulfment of virions into cells (see Sect. 2). The overall lollipop structure of σ1 includes an N-terminal fibrous tail and a C-terminal globular head (FRASER et al. 1990). Dissection of individual protein domains indicated that the N-terminal domain anchors the protein to virions (LEE and LEONE 1993). Not only full-length protein, but N- and C-terminal σ1 fragments can also trimerize (and by distinct mechanisms) (LEONE et al. 1992; STRONG et al. 1991). Trimerization of σ1 is apparently associated with conformational alterations that are essential for cell attachment (LEONE et al. 1991a, b). Folding of σ1 into an active conformation is achieved when the two (N- and C-terminal) trimerization reactions are complete, and receptor binding of σ1 can then be initiated by the interaction of the C-terminal globular head with the receptor(s) (FERNANDES et al. 1994; GILMORE et al. 1996; LEONE et al. 1996). The N-terminal portion of σ1 is an α-helical coiled-coil structure (BASSEL-DUBY et al. 1985). Studies of N-terminal trimerization suggested that it is an ATP-independent, cotranslational process without chaperone involvement (GILMORE et al. 1996). By contrast, C-terminal trimerization requires ATP and Hsp70 and occurs post-translationally (LEONE et al. 1996). The N-terminal trimeric α-helical coiled coil is more stable than the C-terminal trimeric globular head (TURNER et al. 1992). Description of two distinct oligomerization processes in σ1 maturation has provided new insights into nascent protein folding. Trimerization of σ1 is likely an ordered and active process (LEONE et al. 1996) that includes assembly of a loose, triple-coiled coil in the incomplete N-terminal portion of the growing polypeptide chain, with Hsp70 and possibly other chaperones later associating with the C-terminal region to prevent misfolding. Release of the chaperone(s) with ATP hydrolysis subsequently results in globular trimerization of the completed cell attachment protein (LEONE et al. 1996).

6.2 Myristoylation of μ1 and Cleavage to μ1C

Outer-capsid protein μ1C, together with σ3, forms the bulk of the reovirion outer shell. It also plays an important role in virus penetration into host cells. Sequence studies revealed that μ1 contains a conserved N-myristoylation motif at the amino terminus (NIBERT et al. 1991), and analysis of [^{3}H]myristic acid-labeled virus (NIBERT et al. 1991) and M2 cDNA-transfected cells (TILLOTSON and SHATKIN 1992) demonstrated that modified μ1 contains an amide-linked myristoyl group at Gly2, like all other N-myristoylated proteins. The myristoylated 4-kDa peptide μ1N, corresponding to the N-terminal fragment resulting from cleavage of μ1 to μ1C, and intact myristoylated μ1 are both present in virions and ISVP, suggesting that they are also components of the outer shell, in addition to the predominant μ1C protein (NIBERT et al. 1991). Purified mammalian N-myristoyltransferase (NMT) prenylated μ1 in vitro with a K_m of 111 mM (RAJU et al. 1994). Evidence suggests that addition of myristic (or palmitic) acid to capsid or envelope proteins of viruses increases their hydrophobicity and affinity for membranes and may also facilitate protein–protein interactions (HRUBY and FRANKE 1993). An apparent enhancement of membrane association by myristoylated wild-type μ1 relative to the myristoylation-minus G2A mutant is consistent with a function involving membrane interactions during reovirus entry. In addition, myristoylation of μ1 is necessary for conversion of μ1 to μ1C, suggesting that membrane attachment may be important for μ1 proteolytic processing and possibly also for reovirus assembly (TILLOTSON and SHATKIN 1992).

Analysis of virus outer-shell components showed that μ1C is a 72-kDa C-terminal fragment of μ1 and the major M2 gene product in virions. Formation of μ1C by cleavage was originally suggested from the correlated decrease in μ1 and increase in μ1C during infection (LEE et al. 1981; SMITH et al. 1969; ZWEERINK and JOKLIK 1970). Sequence analysis demonstrated that cleavage occurred between Asn-42 and Pro-43 (JAYASURIYA et al. 1988; PETT et al. 1973), and studies of site-specific M2 mutants indicated that scission was sequence specific and myristoylation dependent (TILLOTSON and SHATKIN 1992). The Asn-Pro cleavage site is unusual and apparently has not been reported in any other proteins. In addition, association with σ3 is required for μ1 to μ1C cleavage, at least in cotransfected cells (TILLOTSON and SHATKIN 1992), and processing is efficient in infected cells (LEE et al. 1981b; SMITH et al. 1969), although the origin of the proteolytic activity remains unknown. Since myristoylation and σ3 are necessary, but not sufficient to induce cleavage in vitro, other viral or cellular factors may be required (Z. YUE and A.J. SHATKIN, unpublished results). The proteolytic activity necessary for μ1C formation was absent in lysates of reovirus-infected HeLa cells, although the same lysate retained the capacity to synthesize viral proteins, including full-length μ1 (Z. YUE and A.J. SHATKIN, unpublished results). It is of interest that μ1 contains a conserved GDSG motif that is found in many serine proteases, including chymotrypsin and Sindbis capsid protein (CHOI et al. 1991), and it is possible that μ1 undergoes self-cleavage in a specific cellular compartment where membrane association and/or interactions with σ3 generate conformation(s) that promote the Asn-Pro scission.

Other μ1/μ1C modifications have also been described, including serine or threonine O-glycosylation in some virion μ1C molecules (KRYSTAL et al. 1976), serine phosphorylation in μ1C (KRYSTAL et al. 1975), and polyadenylation of μ1/μ1C (CARTER et al. 1980). However, the significance of these modifications for reovirus multiplication is not clear.

7 Effects of Structural Proteins on Cellular Functions

Because reoviruses productively infect many different types of cells, they are among the most extensively studied models of cytopathogenesis. Reovirus infection usually induces cellular responses that include DNA, RNA, and protein synthesis inhibition; cytoskeleton alterations; disruption of cell membrane integrity; perturbation of signal transduction pathways; and, as reported recently, induction of programmed cell death (apoptosis). Like many other viruses, reoviruses induce interferon and other cytokines that activate host antiviral defense mechanisms, but they also counteract these defenses, notably through the action of the dsRNA-binding, structural protein σ3, which blocks activation of the host PKR (IMANI and JACOBS 1988; LLOYD and SHATKIN 1992).

7.1 Genetic Evidence for Inhibition of DNA and RNA Synthesis by Reovirus Gene Products

Inhibition of DNA synthesis was observed 8–12 h after infection in L cells infected with reovirus type 3 (ENSMINGER and TAMM 1969a; SHARPE and FIELDS 1981). The inhibitory effect was more rapid at higher MOI (SHAW and COX 1973), and it was suggested that the block occurred at G_1 before progression to the S phase of the cell cycle (COX and SHAW 1974; ENSMINGER and TAMM 1969b; HAND et al. 1971). Genetic reassortment indicated that the inhibition of cell DNA synthesis segregated with segment S1 (SHARPE and FIELDS 1981). However, stable or transient expression of type 3 σ1 and σ1NS, either together or individually, did not change the kinetics of DNA replication in transfected cells, although inhibition of DNA synthesis by type 1 infection was enhanced in cells already expressing type 3 σ1 and σ1NS, but not σ1 only (FAJARDO and SHATKIN 1990b). It has been proposed that receptor binding of σ1 provides the signal for inhibition of DNA synthesis based on the finding that treatment of neuroblastoma cell line B104.G4 with an anti-idiotypic, anti-receptor monoclonal antibody resulted in inhibition of DNA synthesis similar to type 3 virus infection (GAULTON and GREENE 1989).

Reovirus infection also inhibits cell RNA synthesis (GOMATOS and TAMM 1963; KUDO and GRAHAM 1966; SHARPE and FIELDS 1982). Gene segment S4, coding for σ3, was reported to segregate with the RNA-inhibitory effect (SHARPE and FIELDS 1982; see also Sect. 7.2), and immunofluorescent analyses of S4-transfected cells demonstrated both nuclear and cytoplasmic localization of σ3 in the absence of other viral proteins (YUE and SHATKIN 1996). Mutant studies indicated that the dsRNA-binding activity of σ3 was correlated with nuclear entry (YUE and SHATKIN 1996). However, it is not known whether the nuclear presence or dsRNA binding of σ3 is related to cell RNA metabolism.

7.2 Role of dsRNA-Binding Protein σ3 in Translational Control

One of the striking consequences of reovirus infection is the shutoff of host protein synthesis that has been observed in many cell types (DANIS and LEMAY 1993; SHARPE and FIELDS 1983; ZWEERINK and JOKLIK 1970). Viral protein synthesis proceeds in the presence of host translational inhibition, which also correlated with accelerated viral multiplication (DANIS and LEMAY 1993; MUNEMITSU and SAMUEL 1984). The inhibition of cellular protein synthesis was mapped to gene segment S4 by reassortment (SHARPE and FIELDS 1982). However, a HeLa cell line that was stably transfected with the type 3 S4 gene produced regulated, functional σ3 protein with no apparent effect on growth or cell division, arguing that σ3 by itself is not sufficient to cause the cytopathic effects, including host translational shutoff, associated with reovirus infections (YUE and SHATKIN 1996).

The mechanism or mechanisms of preferential translation of viral over cellular mRNA are controversial. One series of studies suggested that infection resulted in modification of the cellular translation machinery so that uncapped viral mRNA produced from progeny particles were more efficiently used for translation than capped host mRNA (LEMIEUX et al. 1984; SKUP and MILLWARD 1980b; SKUP et al. 1981). Polypeptide σ3 was shown to participate in translation of late mRNA in infected L cells, implying that this structural protein may be a "factor" that modulates the translation machinery (LEMIEUX et al. 1987). In addition, σ3 was enriched in the translation initiation factor fraction of S4-transfected L cells (LE-MAY and MILLWARD 1986). However, other results indicated that uncapped mRNA were not translated preferentially, suggesting alternatively that abundant viral mRNA successfully outcompeted cell mRNA for limited translational compo-nent(s) (DETJEN et al. 1982; MUNOZ et al. 1985). A translational switch from cellular mRNA to uncapped viral RNA in poliovirus infections is accompanied by cleavage of one of the translation initiation factor polypeptides in the cap-binding complex, but no similar cleavage was detected in reovirus infections (BORMAN et al. 1997; DUNCAN 1990; HENTZE 1997).

More recent studies indicate that σ3 can regulate translation by inhibiting the activation of interferon-induced PKR, an enzyme that shuts off translation initia-tion by phosphorylating the α-subunit of initiation factor eIF-2 (BEATTIE et al. 1995; IMANI and JACOBS 1988; LLOYD and SHATKIN 1992). The 365-amino acid σ3 protein contains two important and separable motifs: a zinc finger at residues 51–71, which is important for stability (MABROUK and LEMAY 1994) and μl binding (SHEPARD et al. 1996), and a conserved dsRNA-binding domain (dsRBD) in the C-terminal half that has a predicted secondary structure similar to other cellular and viral dsRBD (MILLER and SAMUEL 1992; SCHIFF et al. 1988; YUE and SHATKIN 1996). Site-directed mutagenesis of the dsRBD in σ3 implicated several basic residues in dsRNA binding (DENZLER and JACOBS 1994). Binding has been shown by gel mobility shift assays to be both specific for dsRNA and dependent on duplex length (YUE and SHATKIN, 1997). Although it is well documented that activation of PKR by dsRNA (and other treatments) results in inhibition of protein synthesis (CLEMENS 1996) and that the effect in reovirus-infected cells segregates with the S4 gene (SHARPE and FIELDS 1982), the expression of reporter chlo-ramphenicol acetyltransferase (CAT) was not inhibited, but instead enhanced by cotransfection of S4 cDNA (GIANTINI and SHATKIN 1989; LLOYD and SHATKIN 1992). From these results, it was suggested that σ3 can prevent PKR activation by effectively competing for dsRNA activator(s) of PKR. Consistent with this possi-bility, σ3 expressed transiently from transfected S4 cDNA restored viral protein synthesis in cells infected with E3L⁻ mutant vaccinia virus or VAI RNA⁻ mutant dl331 adenovirus (BEATTIE et al. 1995; LLOYD and SHATKIN 1992). Enhancement of reporter and viral protein expression both correlated with the dsRNA-binding capacity of σ3 (BEATTIE et al. 1995; YUE and SHATKIN 1997). Direct competition with PKR for dsRNA binding was also obtained with purified recombinant, ba-culovirus-expressed σ3, and in transfected cells the ability of σ3 to bind dsRNA correlated with PKR inhibition (YUE and SHATKIN 1997). It should be noted that the binding affinity of σ3 for dsRNA is apparently less than PKR, but this dif-ference in infected cells presumably would be compensated for by the presence of large amounts of σ3, a virion structural protein (YUE and SHATKIN 1997).

σ3 exists in complexes with μ1/μ1C, and the presence of μ1/μ1C blocked the dsRNA binding of σ3 (YUE and SHATKIN 1997), probably by altering conformation as shown by changes in protease sensitivity (SHEPARD et al. 1995) and cellular localization (YUE and SHATKIN 1996). Consistent with complex formation and diminished dsRNA binding, μ1/μ1C coexpression in transfected cells prevented the stimulatory effects of σ3 on reporter CAT mRNA translation (TILLOTSON and SHATKIN 1992), demonstrating that μ1/μ1C can regulate σ3 function(s). These findings suggest that σ3 participates in translational control via PKR inhibition early in infection and later forms μ1/μ1C structural complexes required for virus maturation. Consistent with this model, dsRNA-binding activity of σ3 formed early in infection diminished subsequently, probably due to formation of structural complexes (YUE and SHATKIN 1997).

7.3 Reovirus-Induced Apoptosis: σ1 and μ1/μ1C

Cytopathogenesis and cell killing have been extensively studied in reovirus-infected cells. Recent investigations have suggested that reovirus induces apoptosis, as do many other viruses, including human immunodeficiency virus (HIV), adenovirus, and Sindbis virus (SHEN and SHENK 1995). Apoptosis is a host defense mechanism that severely limits virus production and reduces or eliminates the spread of progeny virus in the host (SHEN and SHENK 1995; TEODORO and BRANTON 1997). Reovirus infection can induce morphological hallmarks of apoptosis, including nuclear condensation, blebbing of the plasma membrane, cell shrinkage, and DNA fragmentation into oligonucleosomal ladders in mouse L cells and MDCK canine kidney cells as well as in the central nervous system of the mouse (OBERHAUS et al. 1997; RODGERS et al. 1997; TYLER et al. 1995, 1996). Expression of the apoptosis inhibitor Bcl-2 reduced cell death, but apparently not virus yields in reovirus-infected MDCK cells (RODGERS et al. 1997). Genetic reassortment pointed to σ1 and possibly also μ1/μ1C (TYLER et al. 1995, 1996) as responsible for reovirus-induced apoptosis. Thus early events in infection such as viral attachment and penetration may also initiate programmed cell death in addition to DNA shutoff (TYLER et al. 1995, 1996). Interestingly, dsRNA was recently reported to trigger apoptosis in vaccinia virus-infected cells (KIBLER et al. 1997). This effect is likely mediated by PKR (LEE and ESTEBAN 1994), but the details of how reoviruses and other viruses either trigger or prevent apoptosis are only beginning to be deciphered.

7.4 Other Cytopathic Effects of Structural Proteins

The cytoskeletal architecture is altered in reovirus-infected cells (DALES 1963, 1965; SHARPE et al. 1982), and the reorganization of intermediate filaments and microtubules may contribute to the formation of reovirus replication factories and their movement toward the nucleus with concomitant cell injury (NIBERT et al. 1996).

Although virus assembly may occur on intermediate filaments and viral proteins may associate with microtubules (BABISS et al. 1979; MORA et al. 1987), replication can proceed in cells in which these structures have been disrupted (SPENDLOVE et al. 1964). Reoviruses can also establish nonlytic, persistent infections in a variety of cell types, and the establishment and maintenance of this process has been assigned to viral structural proteins $\sigma3$, $\lambda2$, and $\sigma1$ by reassortant analyses (see the chapter in vol. 2 by Dermody on "Molecular Mechanisms of Persistent Infection by Reovirus").

Acknowledgements. We thank colleagues for communicating results before publication and Ms. Janet Hansen for editorial assistance.

References

Acs G, Klett H, Schonberg M, Christman J, Levin DH, Silverstein SC (1971) Mechanism of reovirus double-stranded RNA synthesis in vivo and in vitro. J Virol 8:684–689

Antczak JB, Chmelo R, Pickup DJ, Joklik WK (1982) Sequences at both termini of the ten genes of reovirus serotype 3 (strain Dearing). Virology 121:307–319

Antczak JB, Joklik WK (1992) Reovirus genome segment assortment into progeny genomes studied by the use of monoclonal antibodies directed against reovirus proteins. Virology 187:760–776

Armstrong GD, Paul RW, Lee PW (1984) Studies on reovirus receptors of L cells: virus binding characteristics and comparison with reovirus receptors of erythrocytes. Virology 138:37–48

Babiss LE, Luftig RB, Weatherbee JA, Weihing RR, Ray UR, Fields BN (1979) Reovirus serotypes 1 and 3 differ in their in vitro association with microtubules. J Virol 30:863–874

Banerjee AK, Shatkin AJ (1970) Transcription in vitro by reovirus-associated ribonucleic acid-dependent polymerase. J Virol 6:1–11

Bartlett JA, Joklik WK (1988) The sequence of the reovirus serotype 3 L3 genome segment which encodes the major core protein $\lambda1$. Virology 167:31–37

Bartlett NM, Gillies SC, Bullivant S, Bellamy AR (1974) Electron microscope study of reovirus reaction cores. J Virol 14:315–326

Bassel-Duby R, Jayasuriya A, Chatterjee D, Sonenberg N, Maizel JV Jr, Fields BN (1985) Sequence of the reovirus haemagglutinin predicts a coiled-coil structure. Nature 315:421–423

Beattie E, Denzler K, Tartaglia J, Perkus M, Paoletti E, Jacobs BL (1995) Reversal of the interferon-sensitive phenotype of a vaccinia virus lacking E3L by expression of the reovirus S4 gene. J Virol 69:499–505

Belli BA, Samuel CE (1993) Biosynthesis of reovirus-specified polypeptides. Identification of regions of the bicistronic reovirus-S1 messenger-RNA that affect the efficiency of translation in animal cells. Virology 193:16–27

Bisaillon M, Bergeron J, Lemay G (1997) Characterization of the nucleoside triphosphate phosphohydrolase and helicase activities of the reovirus $\lambda1$. J Biol Chem 272:18298–18303

Bodkin DK, Fields BN (1989) Growth of reovirus in intestinal tissue: role of the L2 and S1 genes. J Virol 63:1188–1193

Bodkin DK, Nibert MK, Fields BN (1989) Proteolytic digestion of reovirus in the intestinal lumens of neonatal mice. J Virol 63:4676–4681

Borman AM, Kirchweger R, Ziegler E, Rhoads RE, Skern T, Kean KM (1997). eIF4G and its proteolytic cleavage products: effect on initiation of protein synthesis from capped, uncapped, and IRES-containing mRNAs. RNA 3:186–196

Borsa J, Graham AF (1968) Reovirus RNA polymerase activity in purified virions. Biochem Biophys Res Commun 33:895–901

Borsa J, Grover J, Chapman JD (1970) Presence of nucleoside triphosphate phosphohydrolase activity in purified virions of reovirus. J Virol 6:295–302

Borsa J, Copps TP, Sargent MD, Long DG, Chapman JD (1973) New intermediate subviral particles in the in vitro uncoating of reovirus virions by chymotrypsin. J Virol 11:552–564

Borsa J, Long DG, Sargent MD, Copps TP, Chapman JD (1974) Reovirus transcriptase activation in vitro: involvement of an endogenous uncoating activity in the second stage of the process. Intervirology 4:171–188

Borsa J, Sargent MD, Lievaart PA, Copps TP (1981) Reovirus: evidence for a second step in the intracellular uncoating and transcriptase activation process. Virology 111:191–200

Bruck C, Co MS, Slaoui M, Gaulton GN, Smith T, Fields BN, Mullins JI, Greene MI (1986) Nucleic acid sequence of an internal image-bearing monoclonal anti-idiotype and its comparison to the sequence of the external antigen. Proc Natl Acad Sci USA 83:6578–6582

Bruenn JA (1991) Relationships among the positive strand and double-stranded RNA viruses as viewed through their RNA-dependent RNA polymerases. Nucleic Acids Res 19:217–226

Canning WM, Fields BN (1983) Ammonium chloride prevents lytic growth of reovirus and helps to establish persistent infection in mouse L cells. Science 219:987–988

Carter C, Stoltzfus CM, Banerjee AK, Shatkin AJ (1974) Origin of reovirus oligo(A). J Virol 13:1331–1337

Carter C, Lin B, Metley M (1980) Polyadenylation of reovirus proteins. J Biol Chem 255:6479–6485

Chang C-T, Zweerink HJ (1971) Fate of parental reovirus in infected cell. Virology 46:544–555

Chappell JD, Gunn VL, Wetzel JD, Baer GS, Dermody TS (1997) Mutations in type 3 reovirus that determine binding to sialic acid are contained in the fibrous tail domain of viral attachment protein σ1. J Virol 71:1834–1841

Choi AHC (1994) Internalization of virus binding proteins during entry of reovirus into K562 erythroleukemia cells. Virology 200:301–306

Choi H-K, Tong L, Minor W, Dumas P, Boege U, Rossmann MG, Wengler G (1991) Structure of Sindbis virus core protein reveals a chymotrypsin-like serine proteinase and the organization of the virion. Nature 354:37–43

Chow N-L, Shatkin AJ (1975) Blocked and unblocked 5′ termini in reovirus genome RNA. J Virol 15:1057–1064

Clemens MJ (1996) Protein kinases that phosphorylate eIF2 and eIF2B and their role in eukaryotic cell translational control. In: Hershey J, Mathews M, Sonenberg N (eds) Translational control. CSHL, Plainview, NY, pp 139–172

Cleveland DR, Zarbl H, Millward S (1986) Reovirus guanylyltransferase is L2 gene product. J Virol 60:307–311

Co MS, Gaulton GN, Fields BN, Greene MI (1985a) Isolation and biochemical characterization of the mammalian reovirus type 3 cell-surface receptor. Proc Natl Acad Sci USA 82:1494–1498

Co MS, Gaulton GN, Tominaga A, Homcy CJ, Fields BN, Greene MI (1985b) Structural similarities between the mammalian β-adrenergic and reovirus type 3 receptors. Proc Natl Acad Sci USA 82:5315–5318

Coombs KM (1996) Identification and characterization of a double-stranded RNA reovirus temperature-sensitive mutant defective in minor core protein μ2. J Virol 70:4237–4245

Cox DC, Shaw JE (1974) Inhibition of initiation of cellular DNA synthesis after reovirus infection. J Virol 13:760–761

Dales S (1963) Association between the spindle apparatus and reovirus. Proc Natl Acad Sci USA 50:268–275

Dales S (1965) The uptake and development of reovirus in strain L cells followed with labelled viral ribonucleic acid and ferritin-antibody conjugate. Virology 25:193–211

Danis C, Lemay G (1993) Protein synthesis in different cell lines infected with orthoreovirus serotype 3: Inhibition of host-cell protein synthesis correlates with accelerated viral multiplication and cell killing. Biochem Cell Biol 71:81–85

Denzler KL, Jacobs BL (1994) Site-directed mutagenic analysis of reovirus σ3 protein binding to dsRNA. Virology 204:190–199

Dermody TS, Schiff LA, Nibert ML, Coombs KM, Fields BN (1991) The S2 gene nucleotide sequences of prototype strains of the three reovirus serotypes: characterization of reovirus core protein σ2. J Virol 65:5721–5731

Detjen BM, Walden WE, Thach RE (1982) Translational specificity in reovirus-infected mouse fibroblasts. J Biol Chem 257:9855–9860

Drayna D, Fields BN (1982) Activation and characterization of the reovirus transcriptase: genetic analysis. J Virol 41:110–118

Dryden KA, Wang G, Yeager M, Nilbert ML, Coombs KM, Furlong DB, Fields BN, Baker TS (1993) Early steps in reovirus infection are associated with dramatic changes in supermolecular structure and protein conformation: analysis of virion and subviral particles by cryoelectron microscopy and image reconstitution. J Cell Biol 122:1023–1041

Duncan RF (1990) Protein synthesis initiation factor modifications during viral infections: implications for translational control. Electrophoresis 11:219–227

Duncan R, Horne D, Strong JE, Leone G, Pon RT, Yeung MC, Lee PWK (1991) Conformational and functional analysis of the carboxyl-terminal globular head of the reovirus cell attachment protein. Virology 182:810–819

Ensminger WD, Tamm I (1969a) Cellular DNA and protein synthesis in reovirus-infected cells. Virology 39:357–359

Ensminger WD, Tamm I (1969b) The step in cellular DNA synthesis blocked by reovirus infection. Virology 39:935–938

Epstein RL, Powers ML, Rogart RB, Weiner HL (1984) Binding of iodine-125 labelled reovirus to cell surface receptors. Virology 133:46–55

Ernst H, Filipowicz W, Shatkin AJ (1983) Initiation by RNA polymerase II and formation of runoff transcripts containing unblocked and unmethylated 5′ termini. Proc Natl Acad Sci USA 81:2172–2179

Fajardo JE, Shatkin AJ (1990a) Effects of elongation on the translation of a reovirus bicistronic mRNA. Enzyme 44:235–243

Fajardo JE, Shatkin AJ (1990b) Expression of the two reovirus S1 gene products in transfected mammalian cells. Virology 178:223–231

Fausnaugh J, Shatkin AJ (1990) Active site localization in a viral mRNA capping enzyme. J Biol Chem 265:7669–7672

Fernandes J, Tang D, Leone G, Lee PW (1994) Binding of reovirus to receptor leads to conformation changes in viral capsid proteins that are reversible upon virus detachment. J Biol Chem 269:17043–17047

Fields BN, Eagle H (1973) The pH-dependence of reovirus synthesis. Virology 52:581–583

Fraser RDB, Furlong DB, Trus BL, Nibert ML, Fields BN, Steven AC (1990) Molecular structure of the cell attachment protein of reovirus: correlation of computer-processed electron micrographs with sequence-based predictions. J Virol 64:2990–3000

Furlong DB, Nibert ML, Fields BN (1988) Sigma 1 protein of mammalian reoviruses extends from the surfaces of viral particles. J Virol 62:246–256

Furuichi Y, Shatkin AJ (1976) Differential synthesis of blocked and unblocked 5′-termini in reovirus mRNA: effect of pyrophosphate and pyrophosphatase. Proc Natl Acad Sci USA 73:3448–3452

Furuichi Y, Shatkin AJ (1977) 5′-Termini of reovirus mRNA: ability of viral cores to form caps post-transcriptionally. Virology 77:566–578

Furuichi Y, Morgan M, Muthukrishnan S, Shatkin AJ (1975a) Reovirus messenger RNA contains a methylated blocked 5′-terminal structure m^{7}G(5′)ppp(5′)G^{m}pCp. Proc Natl Acad Sci USA 72:362–366

Furuichi Y, Muthukrishnan S, Shatkin AJ (1975b) 5′-Terminal m^{7}G(5′)ppp(5′)G^{m}p in vivo: identification in reovirus genome RNA. Proc Natl Acad Sci USA 72:742–745

Furuichi Y, Muthukrishnan S, Tomasz J, Shatkin AJ (1976) Mechanism of formation of reovirus mRNA 5′-terminal blocked and methylated sequence m^{7}GpppG^{m}pC. J Biol Chem 251:5043–5053

Gaulton GN, Greene MI (1989) Inhibition of cellular DNA synthesis by reovirus occurs through a receptor-linked signaling pathway that is mimicked by antiidiotypic, antireceptor antibody. J Exp Med 169:197–211

Gentsch JR, Hatfield JW (1984) Saturable attachment sites for type 3 mammalian reovirus on murine L cells and human HeLa cells. Virus Res 1:401–414

Giantini M, Shatkin AJ (1989) Stimulation of chloramphenical acetyltransferase mRNA translation by reovirus capsid polypeptide σ3 in cotransfected COS cells. J Virol 63:2415–2421

Gilmore R, Coffey MC, Leone G, McLure K, Lee PWK (1996) Co-translational trimerization of the reovirus cell attachment protein. EMBO J 15:2651–2658

Gomatos PJ (1967) RNA synthesis in reovirus-infected L929 mouse fibroblasts. Proc Natl Acad Sci USA 58:1798–1805

Gomatos PJ, Tamm I (1963) Macromolecular synthesis in reovirus-infected L cells. Biochim Biophys Acta 72:651–653

Hand R, Ensminger WD, Tamm I (1971) Cellular DNA replication in infections with cytocidal RNA viruses. Virology 44:527–536

LIVERPOOL
JOHN MOORES UNIVERSITY
AVRIL ROBARTS LRC
TEL. 0151 231 4022

Hayes EC, Lee PWK, Miller SE, Joklik WK (1981) The interaction of a series of hybridoma IgGs with reovirus particles: demonstration that the core protein λ2 is exposed on the particle surface. Virology 108:147–155

Hazelton PR, Coombs KM (1995) The reovirus mutant tsA279 has temperature-sensitive lesion in the M2 and L2 genes: the M2 gene is associated with decreased viral protein production and blockade in transmembrane transport. Virology 207:46–58

Hentze MW (1997) eIF4G: a multipurpose ribosome adaptor? Science 275:500–501

Hooper JW, Fields BN (1996a) Monoclonal antibodies to reovirus σ1 and μ1 proteins inhibit chromium release from mouse L cells. J Virol 70:672–677

Hooper JW, Fields BN (1996b) Role of the μ1 protein in reovirus stability and capacity to cause chromium release from host cells. J Virol 70:459–467

Hruby DE, Franke CA (1993) Viral acylproteins: greasing the wheels of assembly. Trends Microbiol 1:20–25

Huismans H, Joklik WK (1976) Reovirus-coded polypeptides in infected cells: isolation of two native monomeric polypeptides with affinity for single-stranded and double-stranded RNA, respectively. Virology 70:411–424

Imani F, Jacobs BL (1988) Inhibitory activity for the interferon-induced protein kinase is associated with the reovirus serotype 1 σ3 protein. Proc Natl Acad Sci USA 85:7887–7891

Jayasuriya AK, Nibert ML, Fields BN (1988) Complete nucleotide sequence of the M2 gene segment of reovirus type 3 Dearing and analysis of its protein product μ1. Virology 163:591–602

Joklik WK (1972) Studies on the effect of chymotrypsin on reovirions. Virology 49:700–715

Joklik WK (1981) Structure and function of the reovirus genome. Microbiol Rev 45:483–501

Joklik WK (1995) What reassorts when reovirus genome segments reassort? J Biol Chem 270:4181–4184

Kapuler AM (1970) An extraordinary temperature dependence of the reovirus transcriptase. Biochemistry 9:4453–4457

Kapuler AM, Mendelsohn N, Klett H, Acs G (1970) Four base-specified nucleoside 5'-triphosphatases in the subviral core of reovirus. Nature 225:1209–1213

Kibler KV, Shors T, Perkins KB, Zeman CC, Banaszak MP, Biesterfeldt J, Langland JO, Jacobs BL (1997) Double-stranded RNA is a trigger for apoptosis in vaccinia virus-infected cells. J Virol 71:1992–2003

Koonin EV (1993) Computer-assisted identification of a putative methyltransferase domain in ns5 protein of flaviviruses and λ2 protein of reovirus. J Gen Virol 74:733–740

Kozak M (1977) Nucleotide sequences of 5'-terminal ribosome-protected initiation regions from two reovirus messages. Nature 269:390–394

Kozak M (1981a) Mechanism of mRNA recognition by eukaryotic ribosomes during initiation of protein synthesis. In: Shatkin AJ (ed) Initiation signals in viral gene expression. Current Topics in Microbiology and Immunology, vol. 93. pp 81–123

Kozak M (1981b) Possible role of flanking nucleotides in recognition of AUG initiator codon by eukaryotic ribosomes. Nucleic Acids Res 9:5233–5262

Kozak M (1982) Analysis of ribosome binding sites from the S1 message of reovirus: initiation at the first and second AUG codons. J Mol Biol 156:807–820

Kozak M (1991) Structural features of eucaryotic mRNAs that modulate the initiation of translation. J Biol Chem 266:19867–19870

Kozak M, Shatkin AJ (1977) Sequence of two 5'-terminal ribosomes-protected fragments from reovirus messenger RNAs. J Mol Biol 112:75–96

Kozak M, Shatkin AJ (1978) Migration of 40S ribosomal subunits on messenger RNA in the presence of edeine. J Biol Chem 253:6568–6577

Krystal G, Winn P, Millward S, Sakuma S (1975) Evidence for phosphoproteins in reovirus. Virology 64:505–512

Krystal G, Perrault J, Graham AF (1976) Evidence for a glycoprotein in reovirus. Virology 72:308–321

Kudo H, Graham AF (1966) Selective inhibition of reovirus induced RNA in L cells. Biochem Biophys Res Commun 24:150–155

Langberg SR, Moss B (1981) Post-transcriptional modification of mRNA: purification and characterization of cap I and cap II RNA (nucleoside-2'-)-methyltransferases from HeLa cells. J Biol Chem 256:10054–10061

Lau RY, Van Alstyne D, Berckmans R, Graham AF (1975) Synthesis of reovirus-specific polypeptides in cells pretreated with cycloheximide. J Virol 16:470–478

Lee PWK, Leone G (1993) Reovirus protein σ1: from cell attachment to protein oligomerization and folding mechanisms. Bioessays 16:199–206

Lee PWK, Hayes EC, Joklik WK (1981a) Protein σ1 is the reovirus cell attachment protein. Virology 108:156–163

Lee PWK, Hayes EC, Joklik WK (1981b) Characterization of anti-reovirus immunoglobulin secreted by cloned hybridoma cell lines. Virology 108:134–146

Lee SB, Esteban M (1994) The interferon-induced double-stranded RNA-activated protein kinase induces apoptosis. Virology 199:491–496

Lemay G, Danis C (1994) Reovirus λ1 protein: affinity for double-stranded nucleic acids by a small amino-terminal region of the protein independent from the zinc finger motif. J Gen Virol 75:3261–3266

Lemay G, Millward S (1986) Expression of the cloned S4 gene of reovirus serotype 3 in transformed eucaryotic cells: enrichment of the viral protein in the crude initiation factor fraction. Virus Res 6:133–140

Lemieux R, Zarbl H, Millward S (1984) mRNA discrimination in extracts from uninfected and reovirus-infected L-cells. J Virol 51:215–222

Lemieux R, Lemay G, Millward S (1987) The viral protein sigma 3 participates in translation of late viral mRNA in reovirus-infected L cells. J Virol 61:2472–2479

Leone G, Duncan R, Lee PWK (1991a) Trimerization of the reovirus cell attachment protein (σ1) induces conformational changes in σ1 necessary for its cell-binding function. Virology 184:758–761

Leone G, Mah DCW, Lee PWK (1991b) The incorporation of reovirus cell attachment protein σ1 into virions requires the amino-terminal hydrophobic tail and the adjacent heptad repeat region. Virology 182:346–350

Leone G, Maybaum L, Lee PWK (1992) The reovirus cell attachment protein possesses two independently active trimerization domains: basis of dominant negative effects. Cell 71:479–488

Leone G, Coffey MC, Gilmore R, Duncan R, Maybaum L, Lee PWK (1996) C-terminal trimerization, but not N-terminal trimerization, of the reovirus cell attachment protein is a posttranslational and Hsp70/ATP-dependent process. J Biol Chem 271:8466–8471

Levin DH, Mendelsohn N, Schonberg M, Klett H, Silverstein S, Kapuler AM (1970) Properties of RNA transcriptase in reovirus subviral particles. Proc Natl Acad Sci USA 66:890–897

Liu M, Mattion NM, Estes MK (1992) Rotavirus VP3 expressed in insect cells possesses guanylyltrans-ferase activity. Virology 188:77–84

Lloyd RM, Shatkin AJ (1992) Translation stimulation by reovirus polypeptide σ3: substitution for VAI RNA and inhibition of phosphorylation of the α subunit of eukaryotic initiation factor 2. J Virol 66:6878–6884

Lucia-Jandris P, Hooper JW, Fields BN (1993) Reovirus M2 gene is associated with chromium release from mouse L cells. J Virol 67:5339–5345

Luftig RB, Kilham S, Hay A, Zweerink HJ, Joklik WK (1972) An ultrastructure study of virions and cores of reovirus type 3. Virology 48:170–181

Mabrouk T, Lemay G (1994) Mutations in a CCHC zinc-binding motif of the reovirus σ3 protein decrease its intracellular stability. J Virol 68:5287–5290

Mah DC, Leone G, Jankowski JM, Lee PW (1990) The N-terminal quarter of reovirus cell attachment protein σ1 possesses intrinsic virion-anchoring function. Virology 179:95–103

Mao ZX, Joklik WK (1991) Isolation and enzymatic characterization of protein λ2, the reovirus gu-anylyltransferase. Virology 185:377–386

Maratos-Flier E, Goodman MJ, Murry AH, Kahn CR (1986) Ammonium inhibits processing and cy-totoxicity of reovirus, a nonenveloped virus. J Clin Invest 78:1003–1007

Martines CG, Guinea R, Benavente J, Carrasco L (1996) The entry of reovirus into L cells is dependent on vacuolar proton-ATPase activity. J Virol 70:576–579

Martinez-Costas J, Varela R, Benavente J (1995) Endogenous enzymatic activities of the avian reovirus S1133: identification of the viral capping enzyme. Virology 206:1017–1026

Metcalf P, Cyrklaff M, Adrian M (1991) The 3-dimensional structure of reovirus obtained by cryoelec-tron microscopy. EMBO J 10:3129–3136

Miller JE, Samuel CE (1992) Proteolytic cleavage of the reovirus sigma 3 protein results in enhanced double-stranded RNA-binding activity: identification of a repeated basic amino acid motif within the C-terminal binding region. J Virol 66:5347–5356

Miura K-I, Watanabe K, Sugiura M, Shatkin AJ (1974) The 5'-terminal nucleotide sequences of the double-stranded RNA of human reovirus. Proc Natl Acad Sci USA 71:3979–3983

Mora M, Partin K, Bhatia M, Partin J, Carter C (1987) Association of reovirus proteins with the structural matrix of infected cells. Virology 159:265–277

Morgan EM, Kingsbury DW (1980) Pyridoxal phosphate as a probe of reovirus transcriptase. Bio-chemistry 19:484–489

Morgan EM, Kingsbury DW (1981) Reovirus enzymes that modify messenger RNA are inhibited by perturbation of the lambda proteins. Virology 113:565–572

Morozov SY (1989) A possible relationship of reovirus putative RNA polymerase to polymerase of positive-strand RNA viruses. Nucleic Acids Res 17:5394–5394

Munemitsu SM, Samuel CE (1984) Biosynthesis of reovirus-specified polypeptides. Multiplication rate but not yield of reovirus serotypes 1 and 3 correlates with the level of virus-mediated inhibition of cellular protein synthesis. Virology 136:133–143

Munoz A, Alonso MA, Carrasco L (1985) The regulation of translation in reovirus-infected cells. J Gen Virol 66:2161–2170

Nagata L, Masri SA, Pon RT, Lee PWK (1987) Analysis of functional domains on reovirus cell attachment protein σ1 using cloned S1 gene deletion mutants. Virology 160:162–168

Nakashima K, LaFiandra AJ, Shatkin AJ (1979) Differential dependence of reovirus-associated enzyme activities on genome RNA as determined by psoralen photosensitivity. J Biol Chem 254:8007–8014

Nibert ML, Fields BN (1992) A carboxy-terminal fragment of protein μ1/μ1C is present in infectious subvirion particles of mammalian reoviruses and is proposed to have a role in penetration. J Virol 66:6408–6418

Nibert ML, Dermody TS, Fields BN (1990) Structure of the reovirus cell-attachment protein: a model for the domain organization of σ1. J Virol 64:2976–2989

Nibert ML, Schiff LA, Fields BN (1991) Mammalian reoviruses contain a myristoylated structural protein. J Virol 65:1960–1967

Nibert ML, Schiff LA, Fields BN (1996) Reoviruses and their replication. In: Fields BN, Knipe DM, Howley PM (eds) Fields virology. Lippincott-Raven, Philadelphia, pp 1557–1596

Noble S, Nibert ML (1997) Characterization of an ATPase activity in reovirus cores and its genetic association with core-shell protein λ1. J Virol 71:2182–2191

Nonoyama M, Millward S, Graham AF (1974) Control of transcription of the reovirus genome. Nucleic Acids Res 1:373–385

Oberhaus SM, Smith RL, Clayton GH, Dermody TS, Tyler KL (1997) Reovirus infection and tissue injury in the mouse central nervous system are associated with apoptosis. J Virol 71:2100–2106

Paul RW, Lee PWK (1987) Glycophorin is the reovirus receptor on human erythrocytes. Virology 159:94–101

Paul RW, Choi AHC, Lee PWK (1989) The alpha-anomeric form of sialic acid is the minimal receptor determinant recognized by reovirus. Virology 172:832–835

Pett DM, Vanaman TC, Joklik WK (1973) Studies on the amino- and carboxyl-terminal amino acid sequences of reovirus capsid polypeptides. Virology 52:174–186

Powell KFH, Harvey JD, Bellamy AR (1984) Reovirus RNA transcriptase: evidence for a conformational change during activation of the core particle. Virology 137:1–8

Qiao X, Qiao J, Mindich L (1997) Stoichiometric packaging of the three genomic segments of dsRNA bacteriophage φ6. Proc Natl Acad Sci USA 94:4074–4079

Raju RV, Kalra J, Sharma RK (1994) Purification and properties of bovine spleen N-myristoyl-CoA protein:N-myristoyltransferase. J Biol Chem 269:12080–12083

Rebai N, Almazan G, Wei L, Greene MI, Saragovi HU (1996) A p65/95 neural surface receptor is expressed at the S-G2 phase of the cell cycle and defines distinct populations. Eur J Neurosci 8:273–281

Rodgers SE, Barton ES, Oberhaus SM, Pike B, Gibson CA, Tyler KL, Dermody TS (1997) Reovirus-induced apoptosis of MDCK cells is not linked to viral yield and is blocked by Bcl-2. J Virol 71: 2540–2546

Roner MR, Roner LA, Joklik WK (1993) Translation of reovirus RNA species m1 can initiate at either of the first two in-frame initiation codons. Proc Natl Acad Sci USA 90:8947–8951

Saragovi HU, Bhandoola A, Lemercier MM, Akbar GK, Greene MI (1995) A receptor that subserves reovirus binding can inhibit lymphocyte proliferation triggered by mitogenic signals. DNA Cell Biol 14:653–664

Sawutz DG, Bassel-Duby R, Homcy CJ (1987) High affinity binding of reovirus type 3 to cells that lack beta adrenergic receptor activity. Life Sci 40:399–406

Schiff LA, Fields BN (1990) Reoviruses and their replication. In: Fields B (ed) Virology. Raven, New York, pp 1275–1306

Schiff LA, Nibert ML, Co MS, Brown E, Fields BN (1988) Distinct binding sites for zinc and double-stranded RNA in the reovirus outer capsid protein σ3. Mol Cell Biol 8:273–283

Seliger LS, Zheng K, Shatkin AJ (1987) Complete nucleotide sequence of reovirus L2 gene and deduced amino acid sequence of viral mRNA guanylyltransferase. J Biol Chem 262:16289–16293

Sharpe AH, Fields BN (1981) Reovirus inhibition of cellular DNA synthesis: role of the S1 gene. J Virol 38:389–392

Sharpe AH, Fields BN (1982) Reovirus inhibition of cellular RNA and protein synthesis: role of the S4 gene. Virology 122:381–391

Sharpe AH, Fields BN (1983) Pathogenesis of reovirus infection. In: Joklik W (ed) The reoviridae. Plenum, New York, pp 229–285

Sharpe AH, Chen LB, Fields BN (1982) The interaction of mammalian reoviruses with the cytoskeleton of monkey kidney CV-1 cells. Virology 120:399–411

Shatkin AJ (1974) Methylated messenger RNA synthesis in vitro by purified reovirus. Proc Natl Acad Sci USA 71:3204–3207

Shatkin AJ (1976) Capping of eucaryotic mRNAs. Cell 9:645–653

Shatkin AJ, Kozak M (1983) Biochemical aspects of reovirus transcription and translation. In: Joklik WK (ed) The reoviridae. Plenum, New York, pp 43–54

Shatkin AJ, LaFiandra AJ (1972) Transcription by infectious subviral particles of reovirus. J Virol 10:698–706

Shatkin AJ, Sipe JD (1968) RNA polymerase activity in purified reoviruses. Proc Natl Acad Sci USA 61:1462–1469

Shatkin AJ, Furuichi Y, LaFiandra AJ, Yamakawa M (1983) Initiation of mRNA synthesis and 5′-terminal modification of reovirus transcripts. In: Compans R, Bishop D (eds) Double-stranded RNA viruses. Elsevier, New York, pp 43–54

Shaw JE, Cox DC (1973) Early inhibition of cellular DNA synthesis by high multiplicities of infections and UV-irradiated reovirus. J Virol 12:704–710

Shen Y, Shenk T (1995) Viruses and apoptosis. Curr Opin Genet Dev 5:105–111

Shepard DA, Ehnstrom JG, Schiff LA (1995) Association of reovirus outer capsid protein σ3 and μ1 causes a conformational change that renders σ3 protease sensitive. J Virol 69:8180–8184

Shepard DA, Ehnstrom JG, Skinner PJ, Schiff LA (1996) Mutations in the zinc-binding motif of the reovirus capsid protein σ3 eliminate its ability to associate with capsid protein μ1. J Virol 70:2065–2068

Shuman S, Liu Y, Schwer B (1994) Covalent catalysis in nucleotidyl transfer reactions: essential motifs in Saccharomyces cerevisiae RNA capping enzyme are conserved in Schizosaccharomyces pombe and viral capping enzymes and among polynucleotide ligases. Proc Natl Acad Sci USA 91:12046–12050

Silverstein SC, Dales S (1968) The penetration of reovirus RNA and initiation of its genetic function in L-strain fibroblasts. J Cell Biol 36:197–230

Silverstein SC, Astell C, Levin DH, Schonberg M, Acs G (1972) The mechanisms of reovirus uncoating and gene activation in vivo. Virology 47:797–806

Skehel JE, Joklik WK (1969) Studies on the in vitro transcription of reovirus RNA catalyzed by reovirus cores. Virology 39:822–831

Skup D, Millward S (1980a) mRNA capping enzymes are masked in reovirus progeny subviral particles. J Virol 34:490–496

Skup D, Millward S (1980b) Reovirus-induced modification of cap dependent translation in infected L cells. Proc Natl Acad Sci USA 77:152–156

Skup D, Zarbl H, Millward S (1981) Regulation of translation in L-cells infected with reovirus. J Mol Biol 151:35–55

Smith RE, Zweerink HJ, Joklik WK (1969) Polypeptide components of virions, top component and cores of reovirus type 3. Virology 39:791–798

Spandidos DA, Krystal G, Graham AF (1976) Regulated transcription of the genomes of defective virions and temperature-sensitive mutants of reovirus. J Virol 18:7–19

Spendlove RS, Lennette EH, Chin JN, Knight CO (1964) Effect of antibiotic agents on intracellular reovirus antigen. Cancer Res 24:1826–1833

Starnes MC, Joklik WK (1993) Reovirus protein λ3 is a poly(C)-dependent poly(G) polymerase. Virology 193:356–366

Strong JE, Leone G, Duncan R, Sharma RK, Lee PW (1991) Biochemical and biophysical characterization of the reovirus cell attachment protein σ1: evidence that it is a homotrimer. Virology 184:23–32

Sturzenbecker LJ, Nibert M, Furlong D, Fields BN (1987) Intracellular digestion of reovirus particles requires a low pH and is an essential step in the viral infectious cycle. J Virol 61:2351–2361

Tang D, Strong JE, Lee PWK (1993) Recognition of the epidermal growth factor receptor by reovirus. Virology 197:412–414

Teodoro JG, Branton PE (1997) Regulation of apoptosis by viral gene products. J Virol 71:1739–1746

Tillotson L, Shatkin AJ (1992) Reovirus polypeptide σ3 and N-terminal myristoylation of polypeptide μ1 are required for site-specific cleavage to μ1C in transfected cells. J Virol 66:2180–2186

Tosteson MT, Nibert ML, Fields BN (1993) Ion channels induced in lipid bilayers by subvirion particles of the nonenveloped mammalian reoviruses. Proc Natl Acad Sci USA 90:10549–10552

Turner DL, Duncan R, Lee PWK (1992) Site-directed mutagenesis of the C-terminal portion of reovirus protein σ1. Evidence for a conformation-dependent receptor-binding domain. Virology 186:219–227

Tyler KL, Squier MKT, Rodgers SE, Schneider BE, Oberhaus SM, Grdina TA, Cohen JJ, Dermody TS (1995) Differences in the capacity of reovirus strains to induce apoptosis are determined by the viral attachment protein σ1. J Virol 69:6972–6979

Tyler KL, Squier MKT, Brown AL, Pike B, Willis D, Oberhaus SM, Dermody TS, Cohen JJ (1996) Linkage between reovirus-induced apoptosis and inhibition of cellular DNA synthesis: role of the S1 and M2 genes. J Virol 70:7984–7991

Verdin EM, King GL, Maratos-Flier E (1989) Characterization of a common high affinity receptor for reovirus serotypes 1 and 3 on endothelial cells. J Virol 63:1318–1325

Virgin HW IV, Mann MA, Fields BN, Tyler KL (1991) Monoclonal antibodies to reovirus reveal structure/function relationship between capsid proteins and genetics of susceptibility to antibody action. J Virol 65:6772–6781

Watanabe Y, Millward S, Graham A (1968) Regulation of transcription of the reovirus genome. J Mol Biol 36:107–123

Weiner DB, Girard K, Williams WV, McPhillips T, Rubin DH (1988) Reovirus type 1 and type 3 differ in their binding to isolated intestinal epithelial cells. Microb Pathog 5:29–40

Weiner HL, Drayna D, Averill DR Jr, Fields BN (1977) Molecular basis of reovirus virulence: role of the S1 gene. Proc Natl Acad Sci USA 74:5744–5748

Weiner HL, Ault KA, Fields BN (1980) Interaction of reovirus with cell surface receptors. I. Murine and human lymphocytes have a receptor for the hemagglutinin of reovirus type 3. J Immunol 124:2143–2148

Wessner DR, Fields BN (1993) Isolation and genetic characterization of ethanol-resistant reovirus mutant. J Virol 67:2442–2447

White CK, Zweerink HJ (1976) Studies on the structure of reovirus cores: selective removal of polypeptide λ2. Virology 70:171–180

Williams WV, Kieber-Emmons T, Weiner DB, Rubin DH, Greene MI (1991) Contact residues and predicted structure of the reovirus type 3 receptor interaction. J Biol Chem 266:9241–9250

Xu P, Miller SE, Joklik WK (1993) Generation of reovirus core-like particles in cells infected with hybrid vaccinia viruses that express genome segments L1, L2, L3, and S2. Virology 197:726–731

Yamakawa M, Furuichi Y, Nakashima K, LaFiandra AJ, Shatkin AJ (1981) Excess synthesis of viral mRNA 5′-terminal oligonucleotides by reovirus transcriptase. J Biol Chem 256:6507–6514

Yamakawa M, Furuichi Y, Shatkin AJ (1982) Reovirus transcriptase and capping enzymes are active in intact virions. Virology 118:157–168

Yeung MC, Gill MJ, Alibhai SS, Shahrabadi MS, Lee PWK (1987) Purification and characterization of the reovirus cell attachment protein σ1. Virology 156:377–385

Yeung MC, Lim D, Duncan R, Shahrabadi MS, Cashdollar LW, Lee PWK (1989) The cell attachment proteins of type 1 and type 3 reovirus are differentially susceptible to trypsin and chymotrypsin. Virology 170:62–70

Yin P, Cheang M, Coombs KM (1996) The M1 gene is associated with difference in the temperature optimum of the transcriptase activity in reovirus core particles. J Virol 70:1223–1227

Yue Z, Shatkin AJ (1996) Regulated, stable expression and nuclear presence of reovirus double-stranded RNA-binding protein σ3 in HeLa cells. J Virol 70:3497–3501

Yue Z, Shatkin AJ (1997) Double-stranded RNA-dependent protein kinase (PKR) is regulated by reovirus structural proteins. Virology 234:364–371

Zou S, Brown EG (1992) Identification of sequence elements containing signals for replication and encapsidation of the reovirus M1 genome segment. Virology 186:377–388

Zweerink HJ, Joklik WK (1970) Studies on the intracellular synthesis of reovirus-specified proteins. Virology 41:501–518

Assembly of the Reovirus Genome

W.K. Joklik

1 Introduction

During the early days of molecular virology, the presumed monomolecular nature of their genomes was considered to be one of the primary virtues of viruses. However, in 1969, irrefutable evidence surfaced that reovirus genomes consist not of one, but of ten molecules of double-stranded (ds)RNA (SHATKIN et al. 1968). Further, it quickly became apparent that this was not a case of a "headful" mechanism at work, such as appears to operate for influenza virus, where virus particles probably contain random 11-segment collections of the eight actual influenza genome segment species, so that roughly one in 25 virus particles contains at least one of each and is therefore infectious (LAMB and CHOPPIN 1983; ENAMI et al. 1991). By contrast, the assembly of genome segments to form the reovirus genome is an extraordinarily efficient and precise process because the ratio of virus particles to infectious units in carefully handled reovirus preparations is essentially 1 (1.6 or less) (SPENDLOVE et al. 1970; LARSON et al. 1994).

What processes control the assembly of reovirus genomes, and what processes ensure that each genome contains one, and one only, of each of the ten distinct

Department of Microbiology, Duke University Medical Center, Durham, NC 27710, USA species of dsRNA? In order to answer this question, we must first discover the stage at which morphogenesis assortment proceeds. Reovirus adsorbs to cells via glycoprotein receptors, which are widely distributed on mammalian cells (PAUL and LEE 1987; CHOI et al. 1990; STRONG et al. 1993). Following uptake in endoplasmic vesicles, subviral particles are liberated into the cytoplasm; these particles have lost part of their outer capsid shell, and within them the ten genome segments are transcribed into ten ssRNA species that exit via 12 highly characteristic and icosahedrally distributed hollow projections or spikes and serve as the reovirus messenger RNAs (and are therefore assigned plus polarity) (CHANG and ZWEERINK 1971; ASTELL et al. 1972; SHATKIN and LAFIANDRA 1972; SILVERSTEIN et al. 1972; BORSA et al. 1981; POWELL et al. 1984; STURZENBECKER et al. 1987). Very soon after they have been transcribed, these ssRNA molecules associate with three reovirus proteins: the ssRNA-binding nonstructural proteins μNS and σNS and the dsRNA-binding protein σ3 (ANTCZAK and JOKLIK 1992), the primary specificity of which may not be dsRNA per se, but rather base-paired hairpin loop stems in ssRNA. The resulting complexes contain one ssRNA molecule and ten to 30 protein molecules, depending on their size.

At some time between 2 and 3 h after infection, depending on parameters such as the multiplicity of infection and temperature, dsRNA molecules begin to be formed, i.e., the plus strands are transcribed into minus strands which remain associated with them. The dsRNA molecules are located in complexes that contain not only proteins μNS, σNS, and σ3, but also λ2 and λ3; significantly, in these complexes the ten dsRNA species are present in strictly equimolar amounts, i.e., they are complete genomes (ANTCZAK and JOKLIK 1992). Thus the assembly, or assortment, of reovirus genomes occurs concomitantly with the transcription of the plus strands into minus strands, which suggests that the protein that controls assortment, i.e., the protein that counts, is protein λ3, the polymerase (ANTCZAK and JOKLIK 1992).

2 Reovirus Genome Segment Reassortment

The key to discovering and characterizing the nature of the molecular interactions involved in the generation of reovirus genomes lies in the availability of systems into which novel genome segments can be introduced. There are two such systems, genome segment reassortment and the infectious reovirus RNA system. Genome segment reassortment occurs when cells are infected with two species of reovirus particles (FIELDS and JOKLIK 1969; FIELDS 1973; CROSS and FIELDS 1976), e.g., particles of different serotypes. There are three reovirus serotypes, ST1, ST2, and ST3. Reovirus serotype is controlled by the nature of protein σ1, i.e., in nature, there are three forms of protein σ1 (and within each form, no doubt, closely related variants) that are serologically distinct in the sense that antibodies against any one do not interact, to any significant extent, with the other two (WEINER et al. 1977,

1980). Interestingly, each of the three forms of genome segment S1, which encodes protein σ1, is associated in nature with its own set of nine other genome segments, the cognate members of which are much more closely related than the three S1 genome segments; in each case, the ST1 and ST3 genome segment pairs are more closely related to each other than to the ST2 genome segment (WIENER and JOKLIK 1989; JOKLIK and RONER 1996). In cells infected with viruses belonging to any two of the three serotypes, roughly 15% of the progeny are reassortants the genomes of which contain all possible combinations of parental genome segments in roughly (though not strictly) equal proportions (BROWN et al. 1983); in other words, the genome segments and proteins of all three serotypes can, to all intents and purposes, interact with their heterologous versions as effectively as with their homologous versions. The disclaimer "to all intents and purposes" is necessary because, although there is no difficulty in generating the reassortants, they cannot compete in nature with viruses with homologous genome segment sets, which are the strains isolated in the field (although here, too, it is necessary to point out that very occasionally genome segment variants arise that either themselves, or as a result of the proteins that they encode, interact more effectively with a heterologous than with the homologous genome segment set, which then becomes established as a stable virus strain) (WIENER and JOKLIK 1989; JOKLIK and RONER 1996).

Reovirus genome segment reassortment exhibits two features with interesting implications. First, the three forms of S1 genome segments, which have diverged more than 80% toward randomness (DUNCAN et al. 1990), are accepted into heterotypic genome segment sets as readily as genome segments that are identical in more than 90% of loci; and, orbivirus and rotavirus genome segments are not accepted into reovirus genomes. This indicates the existence of some kind of recognition signals for acceptance. Second, why does the proportion of reassortants among the progeny in mixedly infected cells not exceed 15%? One would imagine that, given the fact that insertion of heterotypic genome segments occurs with equal frequency regardless of serotype and nature of the genome segment, all progeny genomes in mixedly infected cells would be reassortant genome segment sets. The fact that this is not the case indicates the existence of a limiting factor or component.

3 The Infectious Reovirus RNA System

The second system that permits the introduction of novel segments into the reovirus genome is the infectious reovirus RNA system (RONER et al. 1990). If the single-stranded forms of all ten genome segments of, say, ST3 virus are lipofected into cells both in free form and after being translated for 60 min in a cell-free protein-synthesizing system such as a rabbit reticulocyte lysate, and if these cells are then also infected with reovirus ST1 or ST2, infectious ST3 virus is formed. If the genome segments are lipofected in their double-stranded form, the yield is ten times

higher, and if both single-stranded and double-stranded forms are lipofected together, the yield is ten times higher still; in either case, the RNA that is translated
must be ssRNA, not melted dsRNA. In practice, it is simplest to use the system in
its ssRNA configuration.

This system clearly has the potential for inserting any desired heterotypic/
heterologous genome segment into any reovirus genome. Most of the work to be
described concerns the introduction of such genome segments into the genome of
the ST3 strain Dearing, while some involves the introduction of heterotypic genome
segments into the genome of ST1 strain Lang.

4 Nature of the Signals Required for the Introduction of ST1 and ST2 Genome Segments into the ST3 Genome

The first question to be asked when attempting to introduce novel genetic information into the reovirus genome, such as, in the simplest case, heterotypic genome
segments and beyond that genome segments into which foreign genes have been
cloned, concerns the nature and location of the recognition signals on the incoming
genome segments, that is, the signals that are present in all mammalian reovirus
genome segments, but are absent in orbivirus and rotavirus genome segments. In
order to answer this question, the ten reovirus ST3 genome segments were lipofected into mouse L fibroblasts together with the ST2 S1 genome segment, the
introduction of which into the ST3 genome would be easily detectable by the
generation of particles that can be precipitated by antibodies against ST2 virus
rather than ST3 virus. No such particles were ever detected (the only exception
being rare virus particles that contained not only the ST2, but also the ST3 S1
genome segment, i.e., 11 genome segments; however, these particles were genetically
unstable and the ST2 S1 genome segment was lost after two replaquings). Similar
experiments were carried out with several other ST2 genome segments, all with
negative results (JOKLIK and RONER 1995, 1996; RONER et al. 1995).

This experiment was then carried out in a slightly different form. We had
previously found that the amount of ST3 virus formed was not changed detectably
if the ssRNA population lipofected into cells was first separated into its l, m, and s
size classes by density gradient fractionation, followed by mixing. This permits
testing the infectivity of mixtures of RNA size classes derived from different reovirus serotypes. However, mixtures of RNA size classes derived from members of
different serotypes failed to generate infectious intertypic reassortants, a result that
was in line with that described above (RONER et al. 1995). It was concluded that the
infectious reovirus RNA system, in the configuration described thus far and in
contrast to mixed infection of cells with infectious parents, is incapable of intro-
ducing heterotypic genome segments into the genome of ST3 virus.

It was then decided that, rather than attempting to introduce heterotypic genome segments into the ST3 virus genome, thereby generating sequence reassort-

ants, it might be possible to replace truncated genome segments in defective interfering (DI) particles with the homologous or heterotypic full-length genome segments. It is well known that after repeated passage of reovirus ST3 at high multiplicity, DI particles are formed that lack an intact L1 genome segment (NONOYAMA and GRAHAM 1970; SCHUERCH et al. 1974; SPANDIDOS and GRAHAM 1975), containing in its place a truncated L1 genome segment that lacks its central 2000 bp; furthermore, the proportion of such DI particles among the progeny increases until the number of intact genome-possessing (i.e., wild-type, wt) particles present in each cell (assuming a maximum multiplicity of 100 virus particles per cell, beyond which yields tend to decrease) is too small for the particles to fulfill their helper function of generating adequate amounts of intact protein $\lambda 3$, which generally occurs at a wt to DI particle ratio of somewhat less than 0.01 (i.e., the progeny virus population is more than 99% DI).

Attempts to introduce l1 genome segments into DI genomes by lipofecting DI RNA together with homologous or heterotypic l size class RNA species were uniformly unsuccessful (RONER et al. 1995). This suggested the presence of deleterious mutations in DI M or S size class species. Lipofection of wt l and s size class species together with DI virus m size class species yielded infectious ST3 virus, but the combination of wt l and m together with DI s size class RNA species did not. The four s size class genome segments of DI virus genomes were therefore sequenced. The S1, S2, and S3 genome segments contained no mutations, but the S4 genome segment contained two mutations, namely G74 to A (silent) and G624 to A (E to K). The presence of these two mutations has two very important consequences. First, they permit the introduction into the ST3 genome of the truncated l1 genome segment, a segment that lacks no less than approximately 2000 residues; in other words, it permits the formation of size reassortants. Second, the S4 genome segment that possesses these two mutations is incompatible with the other nine wt genome segments in the sense that virus particles that possess genomes containing it are noninfectious (JOKLIK and RONER 1995, 1996; RONER et al. 1995).

The question thus arises as to whether the presence of these two mutations at G74 and G624 in the S4 genome segment are also essential for the formation of sequence reassortants, i.e., for the introduction of heterotypic (ST1 and ST2) genome segments into the ST3 genome. The three ST3 S size class genome segments of the monoreassortant that contained the ST2 S1 genome were therefore sequenced. No mutations were present in the S2 and S3 genome segments, but the S4 genome segment again contained the G74 and G624 mutations. The same was true for the S2 and S3 9ST3-1ST2 monoreassortants; in all cases, the only mutations found in ST3 genome segments were the two in the S4 genome segment, and the ST2 genome segments that had been introduced contained no mutations (RONER et al. 1995). Finally, several multireassortants were also examined, i.e., reassortants that contained two, three, or four ST2 genome segments. Here, rather than sequencing the S4 genome segments, advantage was taken of the fact that the G624 to A mutation abolishes an Mnl1 restriction endonuclease site, which permitted detection of the presence of the mutation by testing whether appropriate polymerase chain reaction (PCR) products were cleaved by Mnl1. All multireassortants were found to possess the G624 mutation (RONER et al. 1995). The existence of the G74 mutation in these multireassortants has not been specifically established, but there is every likelihood that it also is present in the ST3 S4 genome segment of multi-reassortants, as it is in sequence monoreassortants and in size reassortants. In conclusion, all evidence indicates that the two mutations, G74 to A and G624 to A, function as acceptance signals, the presence of which is essential for the acceptance of truncated or heterotypic genome segments into the ST3 genome. The presence of the S4 variant in heterotypic reassortants can also be demonstrated using an interesting functional test, namely, testing for the generation of wt virus in cells infected with pairs of heterotypic monoreassortants, as in such cells only one of the two expected reassortants would be expected to be formed. Three monoreassort-ants were used, each containing either an L, an M, or an S size class ST2 genome segment on the ST3 background. All pairs generated double reassortants without difficulty, but no pair generated the other reassortant, namely, wt virus, the reason being that the only S4 genome segments present were those containing the accep-tance signals, which, as pointed out above, do not yield functional genomes in conjunction with the other nine wt genome segments (RONER et al. 1995).

The universal presence of the S4 genome segment variant with the G74 and G624 mutations in all heterotypic reassortants suggested that it should now be possible to insert any heterotypic, and presumably also heterologous, genome segment into the ST3 genome. This proved to be the case (RONER et al. 1995). In the presence of the S4 genome segment isolated from a monoreassortant, but not in its absence, both truncated (from DI particle genomes) and heterotypic (both ST1 and ST2) genome segments can be introduced into the ST3 genome. It should be pointed out that these results include the replacement of the ST3 S1 by the ST2 and ST1 S1 genome segments which have diverged enormously toward randomness (85%, 68%, and 96%/91% in the first, second, and third base codon positions; DUNCAN et al. 1990; JOKLIK and RONER 1996), which indicates that sequence similarity plays little role in the ability to replace genome segments. This opens the door to the development of reovirus as an expression vector, provided, of course, that incoming genome segments containing foreign genes possess appropriate recognition signals (because cognate rotavirus and orbivirus genome segments, for example, cannot be introduced into the reovirus genome).

4.1 When Do Acceptance Signals Arise?

The existence of signals that are essential for accepting sequence and size variants of genome segments into the ST3 genome raises two interesting questions concerning their origin. The first is, when do they arise? Acceptance signals are obviously present with sufficient frequency in "normal" virus stocks, which, it should be pointed out, are often generated by infecting cells at rather high multiplicities so as to ensure high yields. However, the multiplicity can easily be too high; when this happens, DI particles begin to predominate (the acceptance signal-containing S4 genome segment variant is of course essential for the generation of these particles).

The question thus arises as to how soon after cloning and how rapidly such variants are generated. Examination of a series of reovirus ST3 strain Dearing virus stocks at various passage levels – at multiplicities of about 10 plaque-forming units (PFU)/cell – revealed the following. Immediately following plaque isolation, the virus possessed no detectable amounts of the S4 genome segment with the G624 mutation, nor was such virus capable of yielding reassortants when crossed with ST2 virus (less than 1% of the yield). After approximately four serial passages of viral stocks, the fraction of S4 genome segments that contained the G624 mutation was 3%–5%, and the yield of reassortants when crossed with ST2 virus was about 4%; in stocks after ten to 20 passages, the frequency of the G624 mutation-containing S4 genome segment was about 25%, and about 15% of the progeny of mixed infections with ST2 virus were reassortants. The presence of this level of variants is of course never detected in practice, because the only effect of their presence is a lowering of the specific infectivity of the virus stock by 25%. Thus the G624, and presumably also the G74, mutation arises rapidly following cloning; whether this is due to hypermutability at these loci or to some selective advantage of an unknown nature is not known (Roner et al. 1995).

4.2 Are the Acceptance Signals Required in RNA or in Protein?

The second question is whether the mutations are required in the ssRNA or in σ3, the protein encoded by the S4 genome segment. If they are required in the former, one could envisage a function essential for either RNA–RNA or RNA–protein recognition; if they are required in the latter, one could imagine either a function in protein–RNA interaction or a function of protein σ3 itself. A functional role for σ3 in genome segment reassortment would not be unexpected in view of the fact that it associates rapidly with ssRNA molecules after they have been transcribed (Antczak and Joklik 1992), presumably via affinity for possibly specific hairpin loop stems; further, it is known that σ3 is a component of the complexes within which progeny dsRNA genome segments are generated (Antczak and Joklik 1992).

In order to determine whether a functional effect of the G624 to A (aspartic acid to lysine) mutation might be directly observable, an experiment was designed to determine whether wt protein σ3 can interfere with the function of variant protein σ3. Mouse L929 fibroblasts were infected with wt ST3 or ST1 virus, with either of two monoreassortants that possessed ST2 genome segments on a ST3 background (and therefore possessed S4 genome segments with the G74 and G624 mutations) and with two monoreassortants that possessed ST3 genome segments on a ST1 background (Roner et al. 1995). These six infected monolayers were then infected with vaccinia virus capable of expressing either wt ST3 protein σ3 or T7 RNA polymerase, or with both. After 48 h, the yields of the six types of reovirus were measured. The results indicated that overexpression of wt ST3 protein σ3 had no effect on the yield of wt ST3 virus, but that it did inhibit the multiplication of ST1 virus and of the monoreassortants that possessed ST1 S4; in other words, there was heterotypic interference caused by σ3, a phenomenon not previously observed but not entirely unexpected. Very interestingly, however, wt ST3 protein σ3 also interfered with the multiplication of virus that contained the variant ST3 S4 genome segment (RONER et al. 1995). This result strongly suggests that the G624 to A mutation affects a function of protein σ3, most likely a function concerned with interaction with RNA.

5 Construction and Isolation of the Double Mutant tsA/C

We have found that insertion of "foreign" genome segments into the reovirus ST3 genome is greatly facilitated when the accepting population of ssRNA molecules lacks the RNA species that is to be replaced; further, absence of this species also circumvents the necessity of having to select the novel genome since there is only one product. We have devised a method for achieving essentially complete specific removal of individual ssRNA species by annealing them with short complementary oligodeoxyribonucleotides (about 15 residues long) and digesting the products with RNase H. We have used this technique to construct the double temperature-sensitive (ts) mutant derived from tsA201 (with a ts mutation in genome segment M2) and tsC447 (with a ts mutation in genome segment S2) (RONER et al. 1997). This mutant, tsA/C, is extraordinarily sensitive to temperature: the ratio of its plaquing efficiency at 30°C to that at 39°C is 300 times greater than that of its more temperature-sensitive parent, tsC447, in which the ratio is about 1500 times greater than that for wt virus. Even at 30°C, the yield achieved by tsA/C is about 30 times smaller than the yields achieved by its two parents, which grow as well as wt virus; at 39°C, tsA/C is unable to replicate at all. It therefore appears that the effects of its two temperature-sensitive lesions are essentially additive, which is perhaps not surprising, but could not be taken for granted. Interestingly, although it fails to replicate at 37°C, tsA/C elicits the formation of significant amounts of neutralizing antibodies – more than those elicited by ultraviolet (UV)-inactivated virus, though less than those elicited by infectious wt virus (RONER et al. 1997). On the basis of these results, it seems that tsA/C would be an excellent model vaccine strain: as a double mutant, it would be extremely safe; however, since it is only a double mutant, it would closely mimic the behavior of wt virus.

The reason why tsA/C is able to cause significant amounts of antibodies to form is an interesting problem. We have hypothesized (RONER et al. 1997) that the mutant protein μ1 (or μ1C) that it encodes may be less cytotoxic than wt protein μ1, which appears to be very cytotoxic (LUCIA-JANDRIS et al. 1993; HAZELTON and COOMBS 1995); wt protein μ1 is also known to play a role in causing apoptosis, which probably eliminates reovirus-infected cells (TYLER et al. 1996). If that is so, cells infected with tsA/C would survive for longer periods of time than cells infected with wt virus, and the epitopes of the proteins that cause neutralizing antibodies to form would be displayed on their surfaces for longer periods of time than on the surfaces of cells infected with wt virus; they would therefore be able to cause significant amounts of neutralizing antibodies to form without the necessity for multiple cell infection cycles. However that may be, identification of the acceptance signals required for the assembly of heterotypic reassortants, together with construction and isolation of *tsA/C*, has demonstrated the power of infectious reovirus RNA technology for generating any desired reovirus genome, even genomes that cannot be selected.

6 Nature of the Signals Required for the Introduction of ST3 Genome Segments into the ST1 Genome

All results discussed thus far relate to the introduction of ST2 genome segments into the ST3 genome. ST2 virus is not as closely related to ST3 virus as ST1 virus; all available evidence, such as the data on the extent of divergence shown in Table 1, indicates that ST2 and the ST1/ST3 precursor diverged first and that ST1 and ST3 diverged subsequently. These evolutionary relationships are confirmed by the nature and behavior of reassortants among the three serotypes. Reassortants of ST1 and ST3 are stable in the sense that they persist in nature for long periods of time and are even capable of executing genetic sweeps of their parents. This is indicated by the fact that the various cognate genome segments of ST1 and ST3 differ enormously in the divergence of their third base codon positions (from 6% to more than 90% divergence toward randomness), which suggests that the present-day ST1 and ST3 viruses are descendants of monoreassortants that established

Table 1. Extent of divergence toward randomness percentages for ST1 versus ST3 and for ST2 versus ST1 and ST3 for five reovirus genome segments

Genome segment	Base codon position	ST1 versus ST3[a]	ST2 versus ST1/ST3[b]
L3	1	6	32
	2	3	30
M2	1	4	11
	2	2	2
S2	1	6	13
	2	0	3
S3	1	5	25
	2	2	8
S4	1	4	12
	2	1	5

The numbers shown here are the percent difference values multiplied by 1.33 (JOKLIK and RONER 1996). For the ST2 versus ST1/ST3 comparisons, the ST2 versus ST1 and ST2 versus ST3 values were calculated first and then averaged.
[a]The sums of the first and second base codon position divergence toward randomness percentages for ST1 versus ST3 are 25 and 8, respectively.
[b]The sums of the first and second base codon position divergence toward randomness percentages for ST2 versus ST1/ST3 are 93 and 48, respectively.

themselves at very different time points during evolution (WIENER and JOKLIK 1989; JOKLIK and RONER 1996). This interpretation is supported by recent studies in which sequence diversity among S2 and S4 genome segments of random isolates of ST1 and ST3 viruses was examined; it was found that phylogenetic trees based on RNA sequence could be established, most of the differences occurring in third base codon positions (CHAPPELL et al. 1994; KEDL et al. 1995).

By contrast, reassortants of the more distantly related ST3 and ST2 viruses, while arising with equal facility, do not establish themselves under natural conditions, most probably because their proteins, both structural and nonstructural, fit less well, which would result in less stable capsids and/or smaller yields. In fact, the third base codon positions of all ten genome segments of ST3 and ST2 have diverged to a similar extent, namely, from 74% to 91% toward randomness. The fact that divergence is not complete is probably due to not all third base codon position replacements being gratuitous (WIENER and JOKLIK 1989; JOKLIK and RONER 1996). Interestingly, and rather surprisingly, the introduction of ST3 genome segments into the closely related ST1 genome proceeds according to rules that are quite different from those required for the introduction of ST2 genome segments into the more distantly related ST3 genome. For the latter, acceptance signals are required in the S4 genome segment, and the sequence of incoming genome segments is irrelevant. For the former, no acceptance signals are required in the recipient genome, but incoming genome segments are variants of wt genome segments in one to four positions, some mutations being silent and others resulting in amino acid replacements; in all cases, the effect of the mutations is to render the incoming ST3 genome segments more similar to the ST1 genome segments that they replace (Table 2) (RONER et al. 1995).

Thus the assortment patterns, i.e., the acceptance patterns for heterotypic genome segments, are quite different for closely and more distantly related reovirus parents. Experimentally, the ST3-ST2 pattern is readily manageable, whereas the ST1-ST3 pattern is not.

Table 2. Mutations in ST3 genome segments in monoreassortants containing nine ST1 genome segments and one ST3 genome segment

Monoreassortant	Mutations
ST1-ST3 M1	G462 to A (R to Q)
	U985 to G (silent)
	G1129 to U (M to I)
ST1-ST3 S1	C77 to U (A to V)
	G108 to U (V to F)
	U378 to C (silent)
	G504 to A (silent)
ST1-ST3 S2	U702 to C (silent)
ST1-ST3 S3	G699 to A (silent)
ST1-ST3 S4	C383 to U (silent)

From RONER et al. (1995).

7 Conclusion

Recognition of the acceptance signals in the ST3-ST2 system and their identifi-cation, together with recognition of the advantages of completely eliminating the genome segment to be replaced from accepting genome segment populations (thereby obviating the need for progeny selection), have opened up the field of reverse genetics for viruses with segmented dsRNA genomes. It should now be possible to identify assortment, packaging, and transcription initiation signals, to identify the functional domains of reovirus proteins, and to develop reovirus into a highly efficient (because the virus replicates to high yields) and safe (because the virus does not appear to cause disease in humans) expression vector with clinical as well as basic applications. Further, when this technology has been transferred to the orbivirus and rotavirus genera, it should be possible to construct highly efficient and safe vaccine strains for clinically and economically very important human and animal pathogens.

Acknowledgements. The work discussed in this review was supported by grant 5RO1 AI 08909 from the National Institutes of Health and by grants 3233 and 3233A from the Council for Tobacco Research.

References

Antczak JB, Joklik WK (1992) Reovirus genome segment assortment into progeny genomes studied by the use of monoclonal antibodies directed against reovirus proteins. Virology 187:760–776

Astell C, Silverstein SC, Levin DH, Acs G (1972) Regulation of the reovirus RNA transcriptase by a viral capsomere protein. Virology 48:648–654

Borsa J, Sargent MD, Lievaart PA, Copps TP (1981) Reovirus: evidence for a second step in the intracellular uncoating and transcriptase activation process. Virology 111:191–200

Brown EG, Nibert ML, Fields BN (1983) The L2 gene of reovirus serotype 3 controls the capacity to interfere, accumulate deletions and establish persistent infection. In: Compans RW, Bishop DHL (eds) Double-stranded RNA viruses. Elsevier, New York

Chang C-T, Zweerink HJ (1971) Fate of parental reovirus in infected cells. Virology 46:544–555

Chappell JD, Goral MI, Rodgers SE, dePamphilis CW, Dermody TS (1994) Sequence diversity within the reovirus S2 gene: reovirus genes reassort in nature, and their termini are predicted to form a pan-handle motif. J Virol 69:750–759

Choi AHC, Paul RW, Lee PWK (1990) Reovirus binds to multiple plasma membrane proteins of mouse L fibroblasts. Virology 178:316–320

Cross RK, Fields BN (1976) Use of an aberrant polypeptide as a marker in three-factor crosses: further evidence for independent assortment as the mechanism of recombination between temperature-sensitive mutants of reovirus type 3. Virology 74:345–362

Duncan R, Horne D, Cashdollar LW, Joklik WK, Lee PWK (1990) Identification of conserved domains in the cell attachment proteins of the three serotypes of reovirus. Virology 174:399–409

Enami M, Sharma G, Benham C, Palese P (1991) An influenza virus containing nine different RNA segments. Virology 185:291–298

Fields BN (1973) Genetic reassortment of reovirus mutants. In: Fox CF (ed) Virus research. Academic, New York

Fields BN, Joklik WK (1969) Isolation and preliminary genetic and biochemical characterization of temperature-sensitive mutants of reovirus. Virology 37:335–342

Hazelton PR, Coombs KM (1995) The reovirus mutant tsA279 has temperature-sensitive lesions in the M2 and L2 genes: the M2 gene is associated with decreased viral protein production and blockade in membrane transport. Virology 207:46–58

Joklik WK, Roner MR (1995) What reassorts when reovirus genome segments reassort? J Biol Chem 270:4181–4184

Joklik WK, Roner MR (1996) Molecular recognition in the assembly of the segmented reovirus genome. Progr Nucleic Acids Res Mol Biol 53:249–281

Kedl R, Schmechel S, Schiff L (1995) Comparative sequence analysis of the reovirus S4 genes from 13 serotype 1 and serotype 3 field isolates. J Virol 69:552–559

Lamb RA, Choppin PW (1983) The gene structure and replication of influenza virus. Annu Rev Biochem 52:467–506

Larson SM, Antczak JB, Joklik WK (1994) Reovirus exists in the form of 13 particle species that differ in their content of protein σ1. Virology 201:303–311

Lucia-Jandris P, Hooper JW, Fields BN (1993) Reovirus M2 gene is associated with chromium release from mouse L cells. J Virol 67:5339–5345

Nonoyama M, Graham AF (1970) Appearance of defective virions in clones of reovirus. J Virol 6:693–694

Paul RW, Lee PWK (1987) Glycophorin is the reovirus receptor on human erythrocytes. Virology 159:94–101

Powell KFH, Harvey JD, Bellamy AR (1984) Reovirus RNA transcriptase: evidence for a conformational change during activation of the core particle. Virology 137:1–8

Roner MR, Sutphin LA, Joklik WK (1990) Reovirus RNA is infectious. Virology 179:845–852

Roner MR, Lin P-N, Nepluev I, Kong L-J, Joklik WK (1995) Identification of signals required for the insertion of heterologous genome segments into the reovirus genome. Proc Natl Acad Sci USA 92:12362–12366

Roner MR, Nepliouev I, Sherry B, Joklik WK (1997) Construction and characterization of a reovirus double ts mutant. Proc Natl Acad Sci USA 94:6826–6830

Schuerch AR, Matsuhisa T, Joklik WK (1974) Temperature-sensitive mutants of reovirus. VI. Mutant ts447 and ts556 particles that lack either one or two genome RNA segments. Intervirology 3:36–46

Shatkin AJ, LaFiandra AJ (1972) Transcription by infectious subviral particles of reovirus. J Virol 10:698–707

Shatkin AJ, Sipe JD, Loh PC (1968) Separation of 10 reovirus genome segments by polyacrylamide gel electrophoresis. J Virol 2:986–991

Silverstein SC, Astell C, Levin DH, Schonberg M, Acs G (1972) The mechanisms of reovirus uncoating and gene activation in vivo. Virology 47:797–786

Spandidos DA, Graham AF (1975) Complementation between temperature sensitive and deletion mutants of reovirus. J Virol 16:1444–1453

Spendlove RS, McClain ME, Lennette EH (1970) Enhancement of reovirus infectivity by extracellular removal or alteration of the virus capsid by proteolytic enzymes. J Gen Virol 8:83–93

Strong JF, Tang D, Lee PWK (1993) Evidence that the epidermal growth factor receptor on host cells confers reovirus infection efficiency. Virology 197:405–411

Sturzenbecker LJ, Nibert M, Furlong D, Fields BN (1987) Intracellular digestion of reovirus particles requires a low pH and is an essential step in the viral infectious cycle. J Virol 61:2351–2361

Tyler KL, Squire MKT, Brown AL, Pike B, Willis D, Oberhaus SM, Dermody TS, Cohen JJ (1996) Linkage between reovirus-induced apoptosis and inhibition of cellular DNA synthesis: role of the S1 and M2 genes. J Virol 70:7984–7951

Weiner HL, Drayna D, Averill DR Jr, Fields BN (1977) Molecular basis of reovirus virulence: role of the S1 gene. Proc Natl Acad Sci USA 74:5744–5748

Weiner HL, Greene MI, Fields BN (1980) Delayed hypersensitivity in mice infected with reovirus. I. Identification of host and viral gene products responsible for the immune response. J Immunol 125:278–282

Wiener JR, Joklik WK (1989) The sequences of the reovirus serotype 1, 2 and 3 L1 genome segments and analysis of the mode of divergence of the reovirus serotypes. Virology 169:194–203

Temperature-Sensitive Mutants of Reovirus

K.M. Coombs

1 Introduction

One of the ultimate goals of contemporary biological research is to gain a detailed understanding of the mechanisms by which important biological processes are carried out. Mutant strains of eukaryotic cells, bacteria, and viruses have been used as powerful tools to facilitate the understanding of an extremely broad range of

Department of Medical Microbiology, 730 William Avenue, University of Manitoba, Winnipeg, Manitoba, Canada R3E 0W3

This review is dedicated to the memory of Professor Bernard Nathan Fields, friend and mentor, who first generated and described most of the mutants that will be discussed here; who developed most of the important genetic methods that not only allowed elucidation of these mutants, but also paved the way for a better understanding of pathogenesis; and who instilled in his trainees the knowledge and desire to seek out and comprehend the "big picture."

such biologic processes. Among the easiest of such mutants to use are conditionally lethal ones. The two general types of conditionally lethal mutants are temperature-sensitive (*ts*) and nonsense (e.g., amber) mutants. *ts* mutants may be broadly classified as either *temperature sensitive* (or thermosensitive) if elevated temperatures induce the mutant phenotype or *cold sensitive* if reduced temperatures induce the mutant phenotype (discussed in more detail below in Sect. 3). *ts* mutants have been used to study membrane protein biogenesis (BERGMANN 1989), endosomal acidification and antigen processing (McCOY 1990), DNA metabolism (OGASA-WARA et al. 1991), intercellular communication (GOODENOUGH et al. 1996), and protein-folding pathways (MITRAKI and KING 1992). In the field of virology, the use of conditionally lethal *ts* virus mutants has allowed the elucidation of protein-folding pathways (MITRAKI and KING 1992; CARLETON and BROWN 1996), viral pathogenesis (SCHWARTZBERG et al. 1993; SHIKOVA et al. 1993), genomic integration (SCHWARTZBERG et al. 1993; WISKERCHEN and MUESING 1995) and replication (H. CHEN et al. 1994; MILLNS et al. 1994; NAGY et al. 1995), virus morphogenesis and assembly (COMPTON et al. 1990; RIXON et al. 1992; BLACK et al. 1994; ERICSSON et al. 1995), and as sources of attenuated virus for vaccines (MURPHY et al. 1988; PRINGLE 1996). The focus of this review will be to describe current research involving mammalian reovirus *ts* mutants. Some reference will be made to earlier work in this field, but for details of work prior to the early 1980s, the reader is referred to the last major review in this area (RAMIG and FIELDS 1983). Brief mention also will be made of *ts* mutants among other members of the *Reoviridae* family; however, reference to more detailed reviews in these areas will be provided below in Sect. 6.

2 History

Reovirus *ts* mutants were originally generated by chemical mutagenesis of wild-type stocks in two different laboratories (IKEGAMI and GOMATOS 1968; FIELDS and JOKLIK 1969). Ikegami treated stocks of reovirus serotype 3, strain Dearing (T3D), with nitrous acid, nitrosoguanidine, or 5-fluorouracil. The mutagen-treated (and nontreated) stocks then were diluted and plated at 30 °C (chosen as the permissive temperature). Plaques were picked and putative *ts* mutant clones identified on the basis that a *ts* mutant would cause cytopathic effect (CPE) to the same extent as would wild-type (wt) T3D at 30 °C, but the ability of that mutant to produce CPE at 37 °C would be markedly less than the ability of T3D to produce CPE at 37 °C. Based on these criteria, Ikegami identified six *ts* clones, two each from the nitrous acid-treated, 5-fluorouracil-treated, and nontreated samples. Fields also treated stocks of reovirus T3D with nitrous acid, nitrosoguanidine, or proflavin. Cells were then infected with the treated virus stocks, and viral progeny was recovered and plated at 31 °C (chosen as the permissive temperature). After plaques became visible, the plates were lightly stained, the area corresponding to each plaque marked, and plates further incubated at the nonpermissive temperature, chosen as 39 °C. Plates were reexamined a few days later. Plaques that had continued to grow were assumed to represent wild-type clones, whereas plaques that failed to grow after shift to the nonpermissive temperature were harvested as putative *ts* mutant clones. Based on these criteria, Fields initially identified 35 *ts* clones (16 from the nitrous acid-treated, 14 from the nitrosoguanidine-treated, and five from the proflavin-treated stocks), and, on the basis of genetic analysis (described more fully below in Sect. 4.1.), assigned each mutant to one of five groups (designated A, B, C, D, or E; Table 1). Another mutant (clone 556) was subsequently identified from the nitrosoguanidine-treated stocks, shown to represent another group (F), and one of the previously identified mutants was reclassified, resulting in the addition of group G (CROSS and FIELDS 1972). Concurrent experiments had indicated that the reovirus genome consisted of ten discrete segments of dsRNA (SHATKIN et al. 1968). Thus, chemical mutagenesis generated *ts* mutants that accounted for seven of the expected ten groups (Table 1). Mutants corresponding to the remaining three groups were rescued from various other sources later (described in further detail below). Because of the simultaneous appreciation and exploitation of the segmented nature of the viral genome, which was more fully understood because of concurrent studies with the new *ts* mutants (FIELDS 1973), and the powerful genetic manipulation that this segmented genome afforded, the mutants isolated by Fields were more fully utilized in the years that followed.

3 A Working Definition of Temperature Sensitivity

The term "temperature-sensitive," when applied to an organism, implies that the replication of that organism is impaired by alterations in temperature. Although the term can be applied equally well to indicate that either higher or lower temperatures are detrimental, "temperature sensitive" (or "thermosensitive") is now more commonly used to indicate that detrimental effects take place at a slightly elevated temperature; the term "cold sensitive" is usually taken to mean that detrimental effects take place at reduced temperatures. This temperature-sensitive effect may be seen as a result of either or both of two general types of experimental conditions. Usually, cells are infected with a virus clone, and the ability of the virus to replicate in cells incubated at a slightly elevated temperature is then determined. Alternatively, temperature sensitivity may be seen if stocks of the virus are pretreated at elevated temperatures and then the ability of the heated virus to replicate under normal conditions is determined and compared to similarly heated wild-type virus. Examples of temperature-sensitive enhanced loss of infectivity have been reported for a few reovirus *ts* mutants (IKEGAMI and GOMATOS 1968); however, the majority of reovirus *ts* clones tested to date do not show enhanced loss of infectivity when stocks of the virus are preincubated at elevated temperatures and then used to infect cells at the permissive temperature (see, e.g., IKEGAMI and GOMATOS 1968;

Table 1. The reovirus temperature-sensitive mutants[a]

Group	Clone	Isolated[b]	Reference
A	16	Nitrosoguanidine	FIELDS and JOKLIK 1969
	25	Nitrosoguanidine	FIELDS and JOKLIK 1969
	40	Nitrosoguanidine	FIELDS and JOKLIK 1969
	201	Proflavin	FIELDS and JOKLIK 1969
	234	Nitrosoguanidine	FIELDS and JOKLIK 1969
	242	Nitrosoguanidine	FIELDS and JOKLIK 1969
	245	Nitrosoguanidine	FIELDS and JOKLIK 1969
	255	Nitrosoguanidine	FIELDS and JOKLIK 1969
	256	Nitrous acid	FIELDS and JOKLIK 1969
	258	Nitrous acid	FIELDS and JOKLIK 1969
	264	Nitrous acid	FIELDS and JOKLIK 1969
	270	Proflavin	FIELDS and JOKLIK 1969
	279	Proflavin	FIELDS and JOKLIK 1969
	290	Proflavin	FIELDS and JOKLIK 1969
	323	Nitrous acid	FIELDS and JOKLIK 1969
	327	Nitrous acid	FIELDS and JOKLIK 1969
	329	Nitrous acid	FIELDS and JOKLIK 1969
	340	Nitrous acid	FIELDS and JOKLIK 1969
	343	Nitrous acid	FIELDS and JOKLIK 1969
	344	Nitrous acid	FIELDS and JOKLIK 1969
	376	Nitrous acid	FIELDS and JOKLIK 1969
	415	Nitrous acid	FIELDS and JOKLIK 1969
	418	Nitrous acid	FIELDS and JOKLIK 1969
	438	Nitrosoguanidine	FIELDS and JOKLIK 1969
	470	Nitrosoguanidine	FIELDS and JOKLIK 1969
	472	Nitrosoguanidine	FIELDS and JOKLIK 1969
	474	Nitrosoguanidine	FIELDS and JOKLIK 1969
	488	Nitrosoguanidine	FIELDS and JOKLIK 1969
B	271	Proflavin	FIELDS and JOKLIK 1969
	352	Nitrous acid	FIELDS and JOKLIK 1969
	405	Nitrous acid	FIELDS and JOKLIK 1969
	26/6	Rescued pseudorevertant	AHMED et al. 1980b
	23.66	Rescued pseudorevertant	COOMBS et al. 1994
C	**447**	Nitrosoguanidine	FIELDS and JOKLIK 1969
D	357	Nitrous acid	FIELDS and JOKLIK 1969
	585	Nitrosoguanidine	FIELDS and JOKLIK 1969; ITO and JOKLIK 1972a
E	**320**	Nitrous acid	FIELDS and JOKLIK 1969
F	**556**	Nitrosoguanidine	FIELDS and JOKLIK 1969; CROSS and FIELDS 1972
	14	High passage	AHMED et al. 1980a
G	**453**	Nitrosoguanidine	FIELDS and JOKLIK 1969; CROSS and FIELDS 1972
	107	High passage	AHMED et al. 1980a
	9/12	Rescued pseudorevertant	AHMED et al. 1980b
	9/15	Rescued pseudorevertant	AHMED et al. 1980b
	9/18	Rescued pseudorevertant	AHMED et al. 1980b
	26/19	Rescued pseudorevertant	AHMED et al. 1980b
	22	Rescued from DI	AHMED and FIELDS 1981
	7	Rescued from DI	AHMED and FIELDS 1981
	28.22	Rescued pseudorevertant	COOMBS et al. 1994

Table 1 (*contd.*)

Group	Clone	Isolated[b]	Reference
H	33	Rescued pseudorevertant	RAMIG and FIELDS 1979
	176	Rescued pseudorevertant	RAMIG and FIELDS 1979
	219	Rescued pseudorevertant	RAMIG and FIELDS 1979
	225	Rescued pseudorevertant	RAMIG and FIELDS 1979
	300	Rescued pseudorevertant	RAMIG and FIELDS 1979
	100	High passage	AHMED et al. 1980a
	26/2	Rescued pseudorevertant	AHMED et al. 1980b
	26/5	Rescued pseudorevertant	AHMED et al. 1980b
	26/8	Rescued pseudorevertant	AHMED et al. 1980b
	26/11	Rescued pseudorevertant	AHMED et al. 1980b
	26/14	Rescued pseudorevertant	AHMED et al. 1980b
	11.2[c]	Rescued pseudorevertant	COOMBS et al. 1994
I	**138**	Rescued pseudorevertant	RAMIG and FIELDS 1979
J	**128**	Rescued pseudorevertant	RAMIG and FIELDS 1979

Bold indicates prototypic mutant of group.
[a] As originally isolated in references indicated. Mutants are listed chronologically.
[b] Mutant isolated after stocks of reovirus T3D were treated with the indicated mutagen; or rescued from extragenically suppressed pseudorevertants, high passage stocks, or defective interfering (DI) virus stocks.
[c] Proposed new prototypic mutant (see text).

MATSUHISA and JOKLIK 1974). Thus, in describing a *ts* reovirus clone, the implicit understanding is that replication of the mutant virus is impaired by incubating the infected cells at elevated temperatures. Therefore, the majority of reovirus *ts* mutants probably belong to the general group of *ts* folding mutants (for recent reviews, see FANE and KING 1991; PREVELIGE and KING 1993; GORDON and KING 1994). The increase in temperature necessary to demonstrate the *ts* phenotype of impaired replication is usually a function of both the virus and the cell in which the virus replicates. In the case of reovirus, which is usually grown in mouse L929 cell lines, this increase may be 0–3 °C above 37 °C, the temperature at which wild-type reovirus is commonly grown. Temperature sensitivity of a given virus clone can be mathematically expressed as a ratio of the ability of the virus to grow at the nonpermissive temperature (now established as 39 °C or higher for reovirus grown in L929 cells) compared to its ability to grow at the lower permissive temperature (now established as 31 °C or 32 °C). One of the easiest ways to compare temperature effects on viral replication is by comparing the ability of the virus to produce plaques when grown on monolayer cells at the two temperatures, as initially done by both Ikegami and Fields (IKEGAMI and GOMATOS 1968; FIELDS and JOKLIK 1969; Fig. 1a). Dividing the nonpermissive titer by the permissive titer generates an efficiency of plating (EOP) value. When wild-type reovirus (T3D) is grown in cells incubated at the higher temperatures, replication is only marginally reduced as compared to replication at the lower temperature; the resulting EOP ratio typically will be within an order of magnitude of 1.0 (Fig. 1a, b). By contrast, a *ts* clone (e.g., *tsC447*), when examined by the same analysis, will normally generate a value significantly lower than 0.1. As indicated above, the absolute value is dependent upon the virus and cell system employed. In L929 cells, some reovirus *ts* mutants

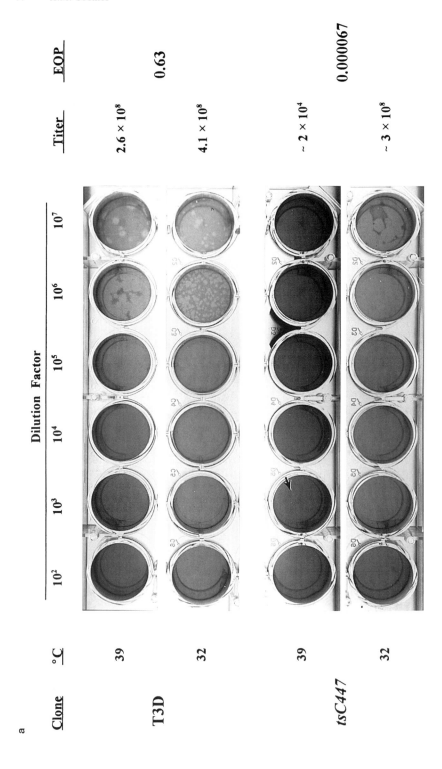

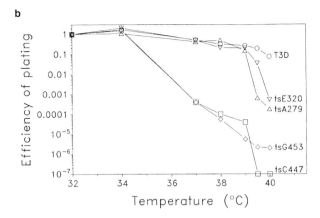

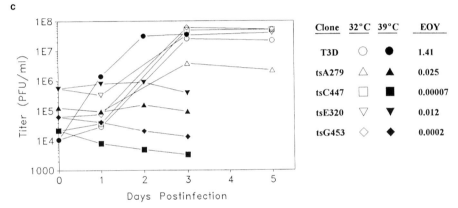

Clone	32°C	39°C	EOY
T3D	○	●	1.41
tsA279	△	▲	0.025
tsC447	□	■	0.00007
tsE320	▽	▼	0.012
tsG453	◇	◆	0.0002

Fig. 1a–c. Expression of temperature sensitivity by reovirus serotype 3 Dearing (T3D) and various temperature-sensitive (*ts*) clones derived from it. **a** Efficiency of plating (EOP) of T3D and *tsC447*. L929 cell monolayers in 24-well dishes were infected with serial tenfold dilutions of each virus stock and the cells incubated at 32 °C and 39 °C. Plates were stained with neutral red and photographed and plaques counted to determine the apparent titer at each temperature. The EOP values (*far right*) were determined by dividing the 39 °C titer by the 32 °C titer for each clone. An example of pinpoint-sized plaques on the 39 °C *tsC447* plate that did not reproduce in the photograph is indicated by the *arrow* in the 10^3 dilution well. **b** Efficiency of plating of T3D and various *ts* mutants at different temperatures. EOP values for each restrictive temperature were determined as described above, except that, for greater precision, virus dilutions were plated in duplicate in six-well plates. EOP values were determined for each temperature by dividing the apparent titer at that temperature by the apparent titer at 32 °C for each clone. Values represent the average of three or more experiments, with experimental variation less than 0.5 $\log_{10}$. **c** Efficiency of yield (EOY) values determined for the same clones as in Fig. 1b. Cells were infected with each virus clone at a multiplicity of infection (MOI) of approximately 0.1 plaque-forming units (PFU) per cell, then incubated at 32 °C and 39 °C. Cells were harvested on the indicated days and disrupted by freeze-thaw and sonication, and the amount of virus present in each sample was determined by plaque assay. EOY values (*far right*) were determined for each clone by dividing the yield of that clone at 39 °C on day 3 by the yield of virus at 32 °C on day 5 (approximately equivalent number of rounds of replication for each temperature) and are tabulated to the *right*

LIVERPOOL JOHN MOORES UNIVERSITY
LEARNING SERVICES

(e.g., *tsC447* and *tsG453*) demonstrate EOP values of 0.0001 or lower at the classical nonpermissive temperature of 39 °C (Fig. 1b). Other clones (e.g., *tsA279* and *tsE320*) show EOP values of approximately 0.1 at this nonpermissive temperature, making it difficult to distinguish such clones from wild type and thus difficult to confidently classify such clones as temperature sensitive.

Alternate methods are available to test the temperature sensitivity of potential *ts* mutants. The plaque assay is only one method to measure viral replication. Because the plaque assay's success requires cell killing, and any number of virions can kill a cell, the relatively simple and rapid EOP assay does not always truly reflect viral replication. An alternative strategy would be to more directly measure viral replication as that amount of virus produced by an infected cell. In such an efficiency of yield (EOY) assay, cells are infected with a virus clone, incubated at various temperatures, and the yield of virus produced at the restrictive temperature is compared to the amount of that virus produced at the permissive temperature. In such an assay, it is advantageous to infect the cells at a relatively low multiplicity of infection (MOI) of less than five plaque-forming units (PFU) per cell and to adjust the incubation times to reflect the different rates of growth at the different temperatures so that progeny viral yield over comparable rounds of replication may be accurately measured. Therefore, it is generally convenient to measure growth at 32 °C after 3–5 days and growth at 39 °C after 2–3 days (approximately three rounds of replication for each). Wild-type virus grows equally well at both temperatures and regularly generates EOY values of approximately 1 (Fig. 1c). Such an assay not only indicates temperature sensitivity of those clones previously identified as *ts* (e.g., both *tsC447* and *tsG453* produce EOY values less than 0.001), but this assay also indicates that the replication of both *tsA279* and *tsE320*, the clones not identified as temperature sensitive by EOP carried out at 39 °C, are restricted in growth more than 20-fold at the higher temperature (Fig. 1c). However, a major disadvantage of the EOY assay is its tediousness; it requires significantly more time and material to perform than an EOP assay. Fortunately, we have found that those *ts* virus clones that did not exhibit temperature-sensitivity when assayed by EOP at 39 °C can be reliably assayed by the simpler and more rapid plaque-plating method if the temperature of incubation is raised slightly higher (HAZELTON and COOMBS 1995; Fig. 1b). At these higher temperatures, EOP values for *ts* mutants may be reduced as much as an additional three orders of magnitude, while the wild-type virus is only marginally affected, thus facilitating identification of *ts* mutant clones.

4 Reovirus Temperature-Sensitive Mutants

4.1 Classification Scheme

The reovirus genome consists of ten segments of dsRNA (SHATKIN et al. 1968; for detailed reviews, see NIBERT et al. 1996b; TYLER and FIELDS 1996). Each of the ten genes may be resolved in polyacrylamide gels, and the resulting pattern of gene migrations is known as the viral electropherotype. Different viral serotypes possess different characteristic electropherotypes (SHARPE et al. 1978). When cells are infected with a mixture of two different serotypes, the genes from the two parents are capable of mixing with each other to generate novel electropherotypes (SHARPE et al. 1978). This process of genetic mixing (called "assortment") generates progeny viruses with genetic mixtures (called "reassortants"; for more detailed reviews, see JOKLIK and RONER 1996; W.K. JOKLIK, this volume.) Because the parental electropherotypes are known, gel analyses of the parents and each reassortant clone allow the parental origin of every gene in any given reassortant to be conveniently and rapidly determined. The segmented nature of these genomes, and the ability to exploit this through reassortant analysis, has profound implications in the use of reovirus in genetic studies such as those described below (for a detailed review, see RAMIG and WARD 1991).

Initial experiments that involved crossing two different ts mutants and scoring the progeny for the production of recombinant wild type-like (ts^+) clones indicated that, in most cases, the ts^+ progeny were generated at either high (approximately 2% or higher) or very low (approximately 0.1% or lower) frequencies (FIELDS and JOKLIK 1969). The absence of a gradient of recombination values in the range 0.1%–2% suggested that recombination was not taking place by the classical means of nucleic acid strand breakage and reunion. Thus, the "all-or-none" values obtained indicated the presence of discrete segments, which, when two different parents were crossed, were free to mix with each other. Similar results and interpretations have been obtained from experiments involving ts mutants of other segmented viruses, such as influenza virus (SIMPSON and HIRST 1968) and rotavirus (GREENBERG et al. 1981; FAULKNER-VALLE et al. 1982; RAMIG 1982).

Reovirus ts mutants were originally classified into groups depending upon whether members of the group, when crossed with other mutant clones, could generate progeny ts^+ virions. While this analysis is functionally equivalent to a standard genetic complementation assay in determining whether any two given clones have their respective ts lesions in the same or different genes, the genetic content of the progeny clones are fundamentally different. Genetic recombination, which, for reovirus, occurs primarily through the process of reassortment (RAMIG and FIELDS 1983), will generate some progeny that stably behaves as wild type; this progeny can be grown under nonpermissive conditions. By contrast, complementation, which occurs through the process of protein mixing, will generate progeny that retains one or the other of the original parental mutations; this progeny will remain incapable of growing under nonpermissive conditions. Several efforts have been made to demonstrate complementation between reovirus mutants (see, e.g., FIELDS and JOKLIK 1969; ITO and JOKLIK 1972a; SPANDIDOS and GRAHAM 1975a, b; CHAKRABORTY et al. 1979). However, this method has not proven reliable for grouping the reovirus ts mutants, possibly because of the phenomenon of genetic interference (CHAKRABORTY et al. 1979; AHMED et al. 1980a; AHMED and FIELDS 1981; ROZINOV and FIELDS 1996) or because the effects are masked by the ability of this virus to genetically reassort. Thus, genetic grouping of the reovirus ts mutants has more commonly been carried out by determining the ability of two different mutants to generate ts^+ progeny during a mixed infection. Recombination is measured by the formula:

$$\frac{(AB)_{NP} - (A + B)_{NP}}{(AB)_P} \times 100$$

where A and B are the two virus clones, AB is the product of the mixed infection, NP is the nonpermissive temperature, and P is the permissive temperature. Two mutants whose lesions reside in the same gene will generate a recombination value of 0 or less (Table 2). Two mutants whose lesions reside in different genes will usually generate a recombination value of 1% or greater. For example, the inability of clone tsA340 to recombine significantly with tsA201, and the concurrent ability to recombine with ts mutants in each of the other nine groups, indicated that clone tsA340 has its ts lesion in the same gene as does tsA201, hence its assignment to the same group (Table 2). Similarly, the inability of clone tsG28.22 to recombine with tsG453 indicated that clone tsG28.22 belongs in the same group as does tsG453. Recombination values of greater than 1% between tsG453 and all the tsA clones, and between tsA201 and tsG28.22, indicate that those designated as group A

Table 2. Recombination between selected reovirus temperature-sensitive mutants

Mutant	Percentage ts^+ recombinants[a] when crossed with:									
	tsA201	tsB352	tsC447	tsD357	tsE320	tsF556	tsG453	tsH26/8	tsI138	tsJ128
tsA201	0[b]	4.2	4.6	11.4	3.1	1.7	2.4	1.4	0.7	0
tsA279	0	6.5	0.9	5.1	2.4	0	5.9	2.0	0.2	0
tsA340	0	3.8	0.4	2.0	0.9	0	1.2	0.4	0	0
tsB271	5.9	0	7.4	5.5	12.3	1.6	0	0.7	0.2	0
tsB352	4.2	0	8.3	10.5	3.6	2.9	3.0	3.6	4.0	0
tsB405	10.8	0	3.0	13.6	3.8	3.3	7.6	0	1.4	4.5
tsC447	–	–	0	15.1	10.2	4.1	7.9	2.7	2.5	0
tsD357	–	–	–	0	12.4	7.4	8.4	10.3	2.6	2.7
tsE320	–	–	–	–	0	2.5	7.6	2.6	0.8	0
tsF556	–	–	–	–	–	0	1.7	5.4	4.2	0
tsG453	–	–	–	–	–	–	0	15.4	1.8	0.2
tsH26/8	–	–	–	–	–	–	–	0	0.3	0
tsI138	–	–	–	–	–	–	–	–	0	0
tsJ128	–	–	–	–	–	–	–	–	–	0
ts11.2	0.3	7.7	0.6	8.6	12.8	5.9	4.9	10.0	1.5	0
ts11.31	0.1	2.8	0.3	11.0	5.0	4.7	0	6.5	0	0
ts23.10	3.6	0.6	21.1	25.6	13.3	32.7	4.5	0	6.8	8.6
ts23.59	0.2	5.9	0	0	0.2	1.1	0.1	3.4	0.2	0.4
ts23.66	1.4	0.1	1.3	0	2.0	0.5	1.0	1.7	2.0	0.6
tsG28.22	1.3	0.9	2.9	2.2	3.0	0.1	0	5.0	0.1	3.6
ts31.13	2.7	1.3	3.2	12.6	8.4	0.9	0.5	7.3	0.1	0
ts31.25	11.3	2.3	0	4.4	3.6	1.2	0.4	0	0.2	0
ts36.27	0	4.2	0.2	0	0	0.6	0	0	0.1	0

Data compiled from FIELDS and JOKLIK 1969, RAMIG and FIELDS 1979, AHMED et al. 1980b, COOMBS 1996.
[a] Percentage ts^+ reassortants = $\{[(\text{titer } AB_{39}) - (\text{titer } A_{39} + \text{titer } B_{39})] \div \text{titer } AB_{32}\} \times 100$.
[b] Calculated value, < 0.05.

members have their lesions in different genes than those designated as group G members. Similar comparisons allowed most of the mutants to be assigned to one of ten groups, corresponding to each of the ten genes (FIELDS and JOKLIK 1969; RAMIG and FIELDS 1979; AHMED et al. 1980a, b; COOMBS 1996). Some *ts* mutants have not been assigned to one of the groups because they appear to contain lesions in more than one gene. For example, clone *ts23.66* appeared to be a double mutant, unable to recombine with mutants in both groups B and D. Clones *ts23.59* and *ts36.27* appeared to be incapable of recombining with prototypic mutants in at least five different groups, suggesting that these clones had many defective genes. Because of the polygenic defects of these later clones, little work has been done on them.

Although the recombination assay can be used as a rapid means to determine whether any two mutants have their lesions in the same or different genes, it is incapable of identifying which mutant gene or genes contain the lesion or lesions. Localization of a *ts* lesion is now determined by genetic reassortment analysis. In earlier mapping experiments, the T3D-derived *ts* mutant of interest was crossed with *ts* mutants derived from other serotypes in which the lesion or lesions of the non-T3D-derived clone were known to be in other genes; ts^+ recombinants were selected and their gene patterns determined, the rationale being that every ts^+ reassortant would have the T3D-derived mutant's *ts* gene replaced by the non-*ts* gene from the other parent. This type of analysis was performed to determine the genomic identity of prototype clones from nine of the ten recombination groups (MUSTOE et al. 1978a; RAMIG et al. 1978; 1983). We have modified and extended the reassortant mapping strategy to also include analysis of the *ts* reassortants generated after crossing a *ts* mutant of interest with the wild-type parent of another serotype. An example of such a mapping analysis, carried out with a clone called tsB405, is shown in Table 3. Several concurrent observations allow unequivocal mapping of the *ts* lesion in *tsB405* to the L2 gene. First, every reassortant with a *ts*-like EOP value (more than 1000-fold lower than the EOP value of T1L, another commonly used wild-type virus) contains the L2 gene derived from the mutant. Second, every reassortant with a ts^+-like EOP value (within an order of magnitude of the EOP value of T1L) contains the L2 gene derived from T1L. Finally, each of the other nine genes is randomly associated with respect to temperature sensitivity. From a practical point of view, genetic mapping may be easier and faster if monoreassortants (reassortant clones that contain one gene derived from one parent and the other nine genes derived from the other parent) are available in the gene of interest. However, it also is advantageous to use many reassortants with unique gene patterns to rule out each of the other nine genes multiple times because of the observation that some reassortants may contain mutated gene segments (JOKLIK and RONER 1995; NIBERT et al. 1996a). Such types of two-panel (*ts* and ts^+ reassortant) analyses have been used to confirm the mapping of many of the groups identified above and to determine the locations of *ts* lesions in newer *ts* clones (COOMBS et al. 1994; HAZELTON and COOMBS 1995; COOMBS 1996; SHING and COOMBS 1996).

Table 3. Electropherotypes and efficiency of plating (EOP) values of T1L × *tsB405* reassortants

Clone	Gene segment										EOP[a]
-	L1	L2	L3	M1	M2	M3	S1	S2	S3	S4	
tsB405	B	**B**	B	B	B	B	B	B	B	B	0.0000082
LB5	1	**B**	1	B	1	B	B	1	1	B	0.000017
LB6	1	**B**	1	B	1	B	1	B	B	B	0.00000058
LB7	1	**B**	1	B	B	B	1	1	B	B	0.00036
LB12	B	**B**	B	B	B	B	1	B	B	B	0.0000015
LB15	B	**B**	1	1	B	B	B	1	B	B	0.0000017
LB18	B	**B**	1	B	B	B	B	B	B	1	0.000012
LB19	B	**B**	B	B	B	B	1	B	B	B	0.0000096
LB27	B	**B**	B	B	1	B	B	B	B	B	0.00014
LB48	B	**B**	B	1	B	B	B	B	B	1	0.0000031
LB62	1	**B**	B	B	B	B	B	B	B	B	0.0000006
T1L	1	**1**	1	1	1	1	1	1	1	1	0.987
LB2	1	**1**	B	B	1	B	B	B	B	B	0.523
LB4	1	**1**	1	B	1	B	B	B	1	1	0.368
LB8	B	**1**	1	1	1	1	B	B	B	1	0.717
LB10	B	**1**	1	1	1	1	1	B	1	B	1.262
LB14	B	**1**	1	B	B	B	1	B	B	B	0.599
Exceptions (*n*)	7	0	6	5	4	3	7	8	4	5	

1, type 1 Lang; B, *tsB405* (parental origin of gene).
[a] Efficiency of plating: ratio of plaque yield at 39 °C to plaque yield at 32 °C.
[b] Bold type indicates gene to which phenotype maps.

The biologic characteristics, affected genes and proteins, representative efficiency of plating values, and the types of particles produced at the nonpermissive temperature of each of the mutants discussed below are summarized in Table 4.

4.2 Group A

Group A is the largest group of reovirus mutants, containing almost 30 different clones (Table 1). All clones were assigned to the same group on the basis that, when mixed infections were analyzed by recombination, as described in the preceding section, these clones did not recombine with each other (FIELDS and JOKLIK 1969). Clone 201 was chosen as the prototypical group A mutant. Clone 201 was one of the five *ts* clones identified after treating T3D with proflavin (Table 1). Thin-section electron microscopy analyses of cells infected with this mutant (or with two other group A mutants, *tsA40* and *tsA270*) at the nonpermissive temperature showed virus inclusions and intact particles that were indistinguishable from those produced by wild-type T3D under normal conditions (FIELDS et al. 1971). Subsequent studies indicated that *tsA201* synthesized normal amounts of protein (FIELDS et al. 1972) and RNA (CROSS and FIELDS 1972) at the nonpermissive temperature. Reassortant mapping experiments showed that this mutant had its *ts* lesion in the M2

Table 4. Characteristics of selected reovirus temperature-sensitive mutants[a]

Group	Clone	Gene	Protein	EOP[b] 39 °C	39.5 °C	40 °C	Synthesis (%)[c] ssRNA	Protein	dsRNA	Particle morphology[d]
A	tsA201	M2	μ1	0.1	0.004	0.002	100	100	100	Normal virions
–	tsA279[e]	M2	μ1	0.2	0.0008	0.0002	ND	1–5	200	Top component[f]
–		L2	λ2	–	–	–	–	–	–	Spike-less core particles
B	tsB352	L2	λ2	0.002	0.00001	7×10^{-7}	25–50	25	25–50	Core-like particles
–	tsB405			8×10^{-6}	1×10^{-6}	1×10^{-7}	ND	ND	ND	Core-like particles
C	tsC447	S2	σ2	0.00004	1×10^{-7}	1×10^{-7}	5	5–10	0.1	Empty outer shells
D	tsD357	L1	λ3	0.005	0.00001	3×10^{-7}	5	10–20	0.1	Empty shells
E	tsE320	S3	σNS	0.25	0.04	0.0006	5	5–10	1	None
F	tsF556	(M3)[g]	μNS	0.08	0.003	0.0002	50–100	50–100	50–100	?
G	tsG453	S4	σ3	5×10^{-6}	2×10^{-6}	2×10^{-6}	20	25	15–25	Core-like particles
H	tsH11.2	M1	μ2	0.0003	0.00006	0.00001	50–100	1	0.1	None
I	tsI138	L3	λ1	0.001	0.00001	5×10^{-6}	ND	50–75	<1	None?
J	tsJ128	S1	σ1	0.3	0.2	0.07	ND	ND	ND	?

ND, not determined.

[a] Selected mutants whose lesions have been mapped and/or which have been studied (see text).
[b] Efficiency of plating at the indicated nonpermissive temperature.
[c] Percentage synthesis of indicated component at nonpermissive temperature compared to synthesis at permissive temperature; compiled from Cross and Fields 1972; Fields et al. 1972; Ito and Joklik 1972a; Coombs 1996; P.R. Hazelton and K.M. Coombs, in preparation.
[d] At the nonpermissive temperature.
[e] Morphology induced by the indicated gene lesion (see text).
[f] Genome-deficient particle with both protein capsids.
[g] Assumed, not definitively mapped (see text).

gene (Mustoe et al. 1978a), which encodes protein µ1 (McCrae and Joklik 1978; Mustoe et al. 1978b).

Shortly after its synthesis, the 708-amino acid (76-kDa) µ1 protein (Jayasuriya et al. 1988; Wiener and Joklik 1988) associates with viral protein σ3 (Lee et al. 1981a). This association causes µ1 to be cleaved to a 666-amino acid peptide designated µ1C (Zweerink and Joklik 1970; Tillotson and Shatkin 1992). Protein µ1C is a major component of the reovirus outer capsid (Fig. 2). Virions contain 600 copies of µ1C (Jayasuriya et al. 1988; Metcalf et al. 1991; Dryden et al. 1993; Nibert et al. 1996b), which accounts for about two thirds of the total protein mass of the outer capsid and about half the total protein mass of the virus. The protein has many important functions. It confers environmental stability to the virus (Drayna and Fields 1982b, c). In addition, it may be important in virus entry; the protein must be proteolytically processed to the δ peptide to allow the infectious subviral particle (ISVP) to penetrate cell membranes (Sturzenbecker et al. 1987; Nibert and Fields 1992; Lucia-Jandris et al. 1993; Tosteson et al. 1993; Nibert et al. 1996b). The group A *ts* mutants have their lesions in one of the most abundant and potentially important structural proteins. Since many other reovirus *ts* mutants are defective in assembly (see below), we decided to examine other group A mutants in the hope of identifying novel phenotypes which could be studied to better understand reovirus replication and functions of µ1. Another group A mutant we have studied is *tsA279*. This mutant was one of several identified by Fields after treatment of T3D with proflavin. This mutant shows marginal temperature sensitivity at the nonpermissive temperature of 39 °C when

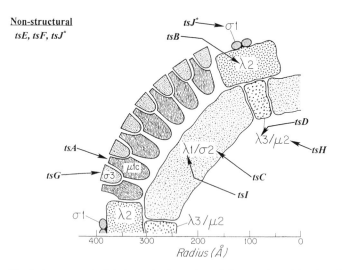

Fig. 2. Arrangement of viral proteins in reovirus. Only a portion of the virus double capsid is shown in cross-section. The presumptive locations of proteins in the outer capsid and inner core are indicated. The specific temperature-sensitive groups, and the structural proteins they affect, are indicated by the italicized letters (*tsA, tsB, ... tsJ*) and *arrows*. Groups that correspond to nonstructural proteins are indicated in the upper left. *Asterisk*, the *tsJ* group corresponds to the S1 gene, which encodes both structural protein σ1 and nonstructural protein σ1s

assayed by EOP (FIELDS and JOKLIK 1969; HAZELTON and COOMBS 1995; see also Fig. 1B). When the nonpermissive temperature was raised only 0.5 °C (to 39.5 °C), *tsA279* more clearly shows a *ts* phenotype (Fig. 1B). When EOP analyses of T1L × *tsA279* reassortants were performed at both 39 °C and 40 °C and the results analyzed, the EOP values fell into three groups (Table 5). When assayed by EOP, using a nonpermissive temperature of 40 °C (EOP_{40}), the members of one group (Table 5, upper panel), which included the parental mutant, all showed EOP_{40} values more than 100-fold lower than the T1L EOP_{40} value. The EOP values of most of these clones also declined by more than two orders of magnitude as the nonpermissive temperature was raised from 39 °C to 40 °C. All these clones

Table 5. Electropherotypes and efficiency of plating (EOP) values of T1Lang × *tsA279* reassortants

Clone	Gene segment										EOP[a]	
	L1	L2	L3	M1	M2	M3	S1	S2	S3	S4	39 °C	40 °C
LA279.08	1	A	1	A	**A**	A	A	A	A	A	0.000139	0.000002
LA279.64	A	1	1	A	**A**	A	A	A	1	A	0.000557	0.000032
LA279.75	1	A	1	1	**A**	A	A	1	A	1	0.00338	0.000072
LA279.20	A	A	1	1	**A**	A	1	A	A	1	0.00568	0.000142
LA279.53	A	A	1	A	**A**	A	A	A	A	1	0.0802	0.000158
tsA279	A	A	A	A	**A**	A	A	A	A	A	0.194	0.000178
LA279.28	A	A	A	1	**A**	A	1	A	1	1	0.000019	0.000195
LA279.43	A	A	1	A	**A**	1	1	A	1	1	0.0156	0.000218
LA279.10	A	1	1	1	**A**	A	1	A	A	A	0.024	0.000441
LA279.60	1	1	A	1	**A**	A	1	A	1	1	0.0735	0.000507
LA279.55	1	A	A	A	**A**	A	A	A	A	A	0.000035	0.00073
LA279.56	1	1	1	1	**A**	1	1	1	1	1	0.171	0.00406
LA279.41	1	**A**	1	1	**1**	A	A	A	A	A	0.0197	0.00296
LA279.76	A	**A**	1	1	**1**	1	1	1	A	A	0.0199	0.0199
LA279.11	A	**A**	A	A	**1**	A	A	A	A	1	0.063	0.0331
LA279.45	A	**A**	1	A	**1**	A	A	A	A	1	0.0395	0.0362
LA279.39	A	**A**	1	1	**1**	A	1	1	1	1	0.105	0.0432
LA279.31	1	**A**	1	1	**1**	A	1	A	A	1	0.382	0.0465
LA279.05	A	**1**	1	1	**1**	1	1	1	A	A	0.647	0.314
LA279.12	1	**1**	1	A	**1**	1	1	1	1	1	0.728	0.369
LA279.46	1	**1**	1	1	**1**	A	1	1	1	1	0.82	0.447
LA279.06	1	**1**	1	1	**1**	A	1	A	1	1	0.933	0.505
T1L	1	**1**	1	1	**1**	1	1	1	1	1	1.03	0.673
LA279.47	A	**1**	1	1	**1**	A	1	A	A	1	1.92	0.708
LA279.63	1	**1**	A	1	**1**	1	1	A	1	1	1.24	0.713
LA279.67	A	**1**	1	A	**1**	1	1	A	1	A	1.52	0.913
LA279.16	1	**1**	1	1	**1**	1	A	A	A	1	0.852	0.974
LA279.33	1	**1**	1	1	**1**	1	1	1	A	1	0.197	1.03
LA279.09	1	**1**	1	A	**1**	A	1	1	A	1	0.885	1.23
LA279.48	1	**1**	1	A	**1**	1	1	A	1	1	0.701	1.31
LA279.34	A	**1**	1	1	**1**	1	1	A	1	1	0.524	1.42
LA279.73	1	**1**	1	A	**1**	1	1	1	A	1	0.834	1.49

Modified from HAZELTON and COOMBS 1995. Reassortants were ranked according to 40 °C EOP data to facilitate grouping. 1, T1Lang; A, *tsA279* (parental origin of gene).
[a] Ratio of plaque yield at indicated temperature to plaque yield at 32 °C.
[b] Bold type indicates gene to which phenotype maps.

contained the M2 gene derived from the mutant parent. In addition, one *ts* clone (LA279.56) was an M2 monoreassortant; it contained only the M2 gene derived from the *tsA279* parent and the other nine genes derived from the wild-type parent. The members of another group of T1L × *tsA279* reassortants, which included the wild-type parent T1L (Table 5, lower panel), all had EOP values within an order of magnitude of the EOP value of T1L, irrespective of whether the nonpermissive temperature was 39 °C or 40 °C. All these clones contained the M2 gene derived from the T1L parent. The combined analyses of these two panels of reassortants confirmed that *tsA279* was a group A mutant. In addition, another group of T1L × *ts*A279 reassortants (Table 5, middle panel) had intermediate EOP values, and these values declined by less than an order of magnitude as the nonpermissive temperature was raised from 39 °C to 40 °C. All these reassortant clones contained the L2 gene derived from the mutant parent, suggesting that *tsA279* contains its primary *ts* lesion in the M2 gene (hence its placement in group A), but that another *ts* lesion was present in the L2 gene (which encodes core spike protein λ2; see Sect. 4.3). Interestingly, the group A prototypic mutant *tsA201* also has a mutation in the L2 gene, although this mutation appears to be silent (Iто and Joκlik 1972b; Schuerch et al. 1974; Mustoe et al. 1978a). The remainder of this section will focus upon what has been learned about the M2 mutation in *tsA279*.

When cells infected with *tsA279* are incubated at the nonpermissive temperature (≥39.5 °C for this mutant) and examined by thin-section electron microscopy, most of the infected cells show no viral inclusions or progeny particle formation (Hazelton and Coombs 1995). Immunofluorescent and immunoprecipitation experiments revealed virtually no progeny viral protein synthesis at the nonpermissive temperature. Metabolic labeling experiments suggest that *tsA279* synthesizes increased levels of dsRNA at the nonpermissive temperature as compared to T3D (P.R. Hazelton and K.M. Coombs, manuscript in preparation). Electron microscopy analysis of infected cells maintained at the nonpermissive temperature also indicated that, at times sufficiently late in infection to allow secondary rounds of replication to occur, a few (presumably) restrictively assembled particles accumulated in lysosomes (Hazelton and Coombs 1995). This uptake could be blocked by administering anti-reovirus antiserum to the cells after primary adsorption but before progeny virus assembly and secondary rounds of replication. In addition, infection of cells at the nonpermissive temperature with viruses grown at the permissive temperature did not reveal accumulation of the permissively assembled virions in such vesicles. These data suggest that the few progeny mutant virions produced at the nonpermissive temperature fail to cross cell membranes to initiate secondary rounds of replication, which suggests a *ts* defect in the ability of *tsA279* to enter cells. Similar experiments with T1L × *tsA279* reassortant viruses indicated that the decrease in protein production, severe reduction in progeny particle formation, and inability of restrictively grown *tsA279* to cross cell membranes were all conferred by the *ts* lesion or lesions in the mutant M2 gene. The latter observation implies that the restrictively assembled *tsA279* virions contain a misfolded μ1 protein that prevents membrane interaction and/or penetration (Hazelton and Coombs 1995). This premise is consistent with other biochemical and genetic observations concerning the roles of the μl and μlC proteins in membrane interactions (STURZENBECKER et al. 1987; NIBERT and FIELDS 1992; LUCIA-JANDRIS et al. 1993; TOSTESON et al. 1993). NIBERT and FIELDS (1992) have proposed that a region near the carboxyl terminus of μl may form an amphipathic helix and may be involved in membrane interactions. Limited sequence information of this region of the *tsA279* M2 gene reveals many alterations (P.R. HAZELTON and K.M. COOMBS, manuscript in preparation), consistent with such a hypothesis. Studies carried out with selected T1L × *tsA279* reassortants that contain the *ts* M2 gene but not the *ts* L2 gene (clones LA279.10 and LA279.56) also indicate that, while progeny viral protein synthesis and particle formation is severely reduced (see above), there is a significant increase in the proportion of genome-deficient "top component" particles produced by the mutant M2 gene at the nonpermissive temperature (P.R. HAZELTON and K.M. COOMBS, manuscript in preparation). These observations suggest a role for μl in genome packaging. A related idea has been proposed by Joklik and Roner, who compared the types of reassortants produced when cells were mixedly infected with either whole virus or core particles (MOODY and JOKLIK 1989; JOKLIK and RONER 1996). Thus, studies with some of the group A mutants reveal defects in processes that the μl protein is believed to be involved in as well as processes not yet known to be associated with the protein. Given the large number of mutants in this group, it is likely that continued study of the group A mutants will provide us with a richer understanding of the roles of this major outer-capsid protein.

4.3 Group B

The mutations in the prototypic group B mutant *tsB352* (MUSTOE et al. 1978a) and in *tsB405* (see Table 3) have been mapped to the L2 gene, which encodes protein λ2 (MCCRAE and JOKLIK 1978). Virions contain 60 copies of this 1290-amino acid (145-kDa) protein. The protein forms the pentameric core spikes (WHITE and ZWEERINK 1976) that reside at each of the particle's icosahedral vertices (LUFTIG et al. 1972; see Fig. 2). Protein λ2 binds guanosine monophosphate (GMP) (FAUSNAUGH and SHATKIN 1990), possesses guanylyltransferase activity (CLEVELAND et al. 1986; MAO and JOKLIK 1991), and serves to attach the cap structure to the nascent mRNA (SHATKIN 1974; FURUICHI et al. 1975; FURUICHI and SHATKIN 1977) as it is extruded from transcribing cores (BARTLETT et al. 1974). There are six mutants that have defective L2 genes, five of which have been assigned to group B. Three of the five were originally isolated by FIELDS (see Table 1), one (*tsA279*; see Sect. 4.2) actually belongs to group A, but contains an additional mutation in the L2 gene (HAZELTON and COOMBS 1995; see Sect. 4.2), and the two newest members (*ts26/6* and *ts23.66*) were isolated as spontaneous mutants during analyses of reovirus *ts* revertants (AHMED et al. 1980b; COOMBS et al. 1994; COOMBS 1996). The *ts23.66* clone appears to be a double mutant, with lesions in both the λ2 and λ3 (group D) proteins (Table 2). Most clones have EOP values of 10^{-3} or less at temperatures of 39 °C or higher; thus they represent "tight" lesions and are relatively easy to work with. Electron microscopy analyses, either of thin-sectioned cells infected with the three original group B mutants at the nonpermissive temperature (FIELDS et al. 1971), or of gradient-purified particles recovered from infected cells (MORGAN and ZWEERINK 1974), showed core-like particles. Thus, these group B mutants appear to have a defect in their λ2 proteins that prevents condensation of the outer capsid proteins onto nascent core particles. Other studies with the group B prototypic mutant tsB352 indicated that this mutant synthesized almost normal amounts of protein (FIELDS et al. 1972) and RNA (CROSS and FIELDS 1972) at the nonpermissive temperature. Studies with the L2 defect in the double mutant tsA279 show different phenotypes. Because this clone contains two ts lesions, it was necessary to study the nature of the L2 lesion by using selected reassortant clones that contain the tsA279 L2 gene segregated away from the dominant mutant M2 gene (see middle panel of Table 5). Examination of some of these reassortants (clones LA279.11 and LA279.76) indicated that virus clones with this mutated L2 gene synthesize normal amounts of viral protein and RNA and assemble core capsid shells, but that neither the mutant λ2 proteins nor outer capsid proteins are capable of condensing onto the nascent core capsids (P.R. HAZELTON and K.M. COOMBS, manuscript in preparation). Thus, this mutant λ2 protein appears to be defective in an earlier step of assembly, attachment of λ2 to the core capsid. Therefore, of the reovirus group B mutants characterized to date, many are assembly defective. These defects are expressed at a minimum of two distinct steps in the assembly pathway.

4.4 Group C

A single group C mutant (tsC447) has been identified to date. This is the most temperature sensitive of the clones routinely used, having an EOP value of about 10^{-5} at 39 °C and higher. The mutation has been mapped to the S2 gene segment (RAMIG et al. 1978), which encodes core protein σ2 (McCRAE and JOKLIK 1978; MUSTOE et al. 1978b). Virions contain between 120 and 180 copies of this 418-amino acid (47-kDa) protein. Protein σ2 is a major component of the core and, along with the other major core capsid protein λ1, is required for assembly of the core capsid (XU et al. 1993; COOMBS et al. 1994). Protein σ2 may occupy an internal position in the core (WHITE and ZWEERINK 1976) and also binds RNA (SCHIFF et al. 1988; DERMODY et al. 1991). Mutant tsC447 produces empty outer capsid structures (FIELDS et al. 1971; MATSUHISA and JOKLIK 1974; COOMBS et al. 1994) at the nonpermissive temperature. Thus, the mutant appears to be defective in assembly of the core particle. Interestingly, the empty outer capsid structures contain protein λ2 (MATSUHISA and JOKLIK 1974), indicating important λ2–outer capsid protein interactions, as also implied by the tsB studies (see above). The mutant produces reduced amounts (approximately 5%) of ssRNA and protein (CROSS and FIELDS 1972; FIELDS et al. 1972) and undetectable amounts (≤0.1%) of progeny dsRNA (CROSS and FIELDS 1972; ITO and JOKLIK 1972a) at the nonpermissive temperature; thus it is classified as an RNA⁻ mutant. However, the ability of the mutant σ2 protein to bind RNA does not appear to be affected (DERMODY et al. 1991). The sequence of the mutant gene has been determined; it contains three nucleotide alterations as compared to parental T3D. These changes are C to U transitions at nucleotide positions 581 and 986, which lead to predicted alanine to valine replacements at amino acid positions 188 and 323, respectively, and an A to G transition at nucleotide position 1166, which leads to a predicted asparagine to aspartic acid change at position 383 in the amino acid sequence (WIENER et al. 1989b). Analyses of *tsC447* revertants indicated that, in contrast to reversion occurring as a result of extragenic suppression, as is the case for most reovirus reversion events examined (RAMIG and FIELDS 1979; MCPHILLIPS and RAMIG 1984; see also RAMIG, this volume), reversion of the *tsC447* lesion appeared to involve intragenic reversion (COOMBS et al. 1994). These studies also identified the mutant aspartic acid residue at amino acid position 383 as responsible for the inability of *tsC447* to assemble core particles at the nonpermissive temperature. The observation that reversion of *tsC447* involved primarily true reversion has a number of implications in considering the normal role of σ2 in core capsid assembly. Suppression by a mutation that lies in a gene other than the site of the original mutation (extragenic) has been interpreted to mean that the two gene products interact with each other at either a structural and/or functional level (for a review, see RAMIG and WARD 1991). However, the *tsC447* reversion mapping results indicated that no other reovirus proteins appeared capable of correcting the defect present in the *tsC447* σ2 protein. Therefore, the failure of *tsC447* to assemble core particles suggests that the region of σ2 near or at Asp-383 is directly responsible for interactions that lead to core shell assembly. This region is predicted to be exposed on the surface of the protein (DERMODY et al. 1991), where it would be expected to be capable of interacting with other proteins. An alternative explanation is that this point mutation may cause global changes in the σ2 protein of the mutant and that the region or regions responsible for recognition and interactions that lead to assembly are far removed from the actual site of the lesion. However, this type of global perturbation is more commonly observed in deletion mutations rather than in point mutations (CREIGHTON 1990). Thus, the introduction of a mutant charged residue in this region in the σ2 amino acid sequence appears to prevent interaction of this protein with other proteins that would normally lead to inner capsid assembly. The primary way in which this alteration appears to be corrected is by true reversion, which suggests that σ2–σ2 interactions involving the region near residue 383 are important in core capsid assembly.

4.5 Group D

The mutation in the prototypical group D mutant (*tsD357*) has been mapped to the L1 gene segment (RAMIG et al. 1978), which encodes core protein λ3 (MCCRAE and JOKLIK 1978). Protein λ3 is the RNA-dependent RNA polymerase (DRAYNA and FIELDS 1982a; MOROZOV 1989; STARNES and JOKLIK 1993). Virions contain about 12 copies of this 1267-amino acid (142-kDa) protein. The precise location of this protein has not yet been reported. It appears to be an internal component of the core capsid (CASHDOLLAR 1994). Several lines of evidence suggest that the protein may be located near the core vertices (see Fig. 2). These include the presumed cooperative roles played by this RNA-dependent RNA polymerase and the capping functions mediated by the core spike λ2 pentamers. In addition, stoichiometric analyses suggest the presence of 12 copies of λ3 and 12 pentameric λ2 spikes in each particle. Finally, recent reconstructions of empty virion structures examined by electron cryomicroscopy suggest that λ3 may be located at the base of the λ2 spikes (K.A. DRYDEN and M.L. NIBERT, personal communication). $tsD357$ has an EOP value close to 10^{-3} at 39 °C. Electron microscopy of restrictively infected cells revealed a heterogeneous mixture of empty particles (FIELDS et al. 1971). The mutant produces reduced amounts (approximately 10%) of ssRNA and protein (CROSS and FIELDS 1972; FIELDS et al. 1972) and undetectable amounts (≤0.1%) of progeny dsRNA (CROSS and FIELDS 1972; ITO and JOKLIK 1972a) at the nonpermissive temperature; thus it is classified as an RNA$^-$ mutant. Another mutant isolated by FIELDS and JOKLIK (1969) after nitrosoguanidine treatment of T3D (clone 585) was subsequently shown to belong to group D (ITO and JOKLIK 1972a). This mutant also has an RNA$^-$ phenotype (ITO and JOKLIK 1972a). Recent work with these two mutants has not been reported.

4.6 Group E

A single group E mutant ($tsE320$) has been identified to date. This is considered a "leaky" ts clone because its EOP value of about 0.2 at 39 °C is indistinguishable from the EOP value of T3D at 39 °C. The EOP value of this mutant can be reduced approximately 50-fold for each 0.5 °C increase in nonpermissive temperature up to 40 °C. We generally use a nonpermissive temperature of 40 °C for this mutant where the EOP value is less than 10^{-3} (P.R. HAZELTON and K.M. COOMBS, manuscript in preparation). The mutation in $tsE320$ has been mapped to the S3 gene segment (RAMIG et al. 1978), which encodes nonstructural protein σNS (MCCRAE and JOKLIK 1978; MUSTOE et al. 1978b). Infected cells contain large amounts of this 366-amino acid (41-kDa) protein. Protein σNS binds ssRNA (HUISMANS and JOKLIK 1976; GOMATOS et al. 1981). Therefore, a mutant in this protein might be expected to be defective in the conversion of ssRNA to dsRNA and hence in progeny particle packaging or production. However, electron microscopy of restrictively infected cells showed normal viral inclusions with normal virus particles (FIELDS et al. 1971). This discrepancy might be explained by the observation that $tsE320$ has an EOP$_{39}$ value of about 0.2, which is indistinguishable from the EOP$_{39}$ value of T3D. Thus, the electron microscopy studies performed at 39 °C may not have been optimized for demonstrating the nonpermissive phenotype of this mutant. Alternatively, the σNS defect may have been expressed but not related to particle morphology. Later studies showed that the mutant produces reduced amounts (approximately 5%) of ssRNA and protein (CROSS and FIELDS 1972; FIELDS et al. 1972) and very low levels (approximately 1%) of progeny dsRNA (CROSS and FIELDS 1972; ITO and JOKLIK 1972a) at the nonpermissive temperature; thus it is classified as an RNA$^-$ mutant. CROSS and FIELDS (1972) also reported that thin-section electron microscopy of cells infected at the nonpermissive temperature with *tsE320* showed no viral inclusions. The sequence of the mutated S3 gene has been determined (WIENER and JOKLIK 1987); it contains a single transition (U$_{806}$ to C), which results in a predicted change from methionine to threonine at amino acid position 260 in the σNS protein. Recent studies carried out with *tsE320* at the higher nonpermissive temperature of 39.5–40 °C, where the mutant phenotype is more clearly expressed, confirm that the mutant produces virtually undetectable amounts of dsRNA and fails to produce any recognizable viral structures (P.R. HAZELTON and K.M. COOMBS, manuscript in preparation).

4.7 Group F

The prototypic group F mutant (*tsF556*) was isolated as part of the original panel (FIELDS and JOKLIK 1969) and subsequently characterized (CROSS and FIELDS 1972). This mutant has proven difficult to work with. Its EOP value at 39 °C is approximately 0.1. In addition, the *ts* lesion has not been unequivocally mapped to any gene. Some experiments suggested that the lesion resided in the M3 gene (RAMIG and FIELDS 1983), which encodes nonstructural protein μNS (McCRAE and JOKLIK 1978; MUSTOE et al. 1978b). Infected cells contain large amounts of this 719-amino acid (80-kDa) protein, which binds ssRNA (HUISMANS and JOKLIK 1976). The mutant also produced normal amounts of protein (FIELDS et al. 1972) and dsRNA (CROSS and FIELDS 1972). More recently, another potential group F mutant has been isolated from high-passage stocks of T3D (AHMED et al. 1980a) and is reported to be easier to work with (RAMIG and FIELDS 1983); however, it has not been further analyzed.

4.8 Group G

Group G is the second largest group of reovirus *ts* mutants. It is of interest that the two largest groups of mutants (this group and group A, see Sect. 4.2) represent both major outer capsid proteins. The prototypic group G mutant (*tsG453*) was initially identified as a group B mutant in the original panel (FIELDS and JOKLIK 1969) and subsequently reclassified (CROSS and FIELDS 1972). This is one of the most *ts* clones, having an EOP value less than 10^{-5} at temperatures of 39 °C or higher. The *ts* lesion in this mutant was mapped to the S4 gene (MUSTOE et al. 1978a), which encodes outer capsid protein σ3 (McCRAE and JOKLIK 1978; MUSTOE et al. 1978b). Virions contain 600 copies of this 365-amino acid (41-kDa) protein. The protein plays important roles in virion stability (DRAYNA and FIELDS 1982b, c), in shutting down host macromolecular synthesis (SHARPE and FIELDS 1982), and in downregulation of interferon-induced dsRNA-activated protein kinase (IMANI and JACOBS 1988; GIANTINI and SHATKIN 1989). Protein σ3 associates with the other outer capsid protein, μl (LEE et al. 1981a), and appears to mediate the cleavage of μl to μlC (ZWEERINK and JOKLIK 1970; TILLOTSON and SHATKIN 1992). The protein contains a zinc finger motif (GIANTINI et al. 1984; ATWATER et al. 1986; SELIGER et al. 1992) and binds zinc (SCHIFF et al. 1988), which seems to be important both for correct folding (DANIS et al. 1992) and stability (MABROUK and LEMAY 1994) of the protein. Protein σ3 has a separate motif that binds dsRNA (HUISMANS and JOKLIK 1976; SCHIFF et al. 1988; MILLER and SAMUEL 1992; DENZLER and JACOBS 1994; MABROUK et al. 1995). The protein is proteolytically removed from virions during infection to generate the ISVP (for a review, see NIBERT et al. 1996b) and may be artificially added back onto purified ISVP (ASTELL et al. 1972). *tsG453* synthesizes reduced amounts (approximately 20%) of protein (FIELDS et al. 1972; DANIS et al. 1992; SHING and COOMBS 1996) and RNA (CROSS and FIELDS 1972) at the re-strictive temperature. The mutant σ3 protein also appears to bind dsRNA to a higher level than does wild-type σ3 (G. LEMAY, personal communication). Many electron microscopy analyses have shown that this mutant produces core-like particles rather than ISVP-like particles at the restrictive temperature (FIELDS et al. 1971; MORGAN and ZWEERINK 1974; DANIS et al. 1992). The apparently paradox-ical observation that a mutation in σ3 caused the accumulation of core-like par-ticles rather than ISVP-like particles was recently resolved when immuno-coprecipitation experiments indicated that σ3–μl interactions are required for condensation of the outer capsid onto nascent cores and that restrictively grown mutant σ3 protein is misfolded such that it cannot interact with μl to form these prerequisite complexes (SHING and COOMBS 1996). The sequence of the mutated S4 gene has been determined. The mutant gene contains three alterations compared to the parental T3D S4 gene: a C to G transversion at nucleotide position 80, a G to A transition at position 455, and a transition from G to U at position 719 (DANIS et al. 1992; SHING and COOMBS 1996). Each of these nucleotide substitutions results in predicted amino acid changes in the σ3 protein; Asn-16 to Lys, which is a significant change in charge; Met-141 to Ile, and Glu-229 to Asp. The identity of the one or more amino acids responsible for the *ts* phenotype has not been de-termined. However, analyses of a variety of expressed mutated σ3 proteins reveal that the amino-terminally located zinc finger, which lies between residues 46 and 73 (SELIGER et al. 1992), is important for correct folding and stability of the protein (DANIS et al. 1992; MABROUK and LEMAY 1994) and its ability to associate with protein μl (SHEPARD et al. 1996), but may not be important for dsRNA binding (DENZLER and JACOBS 1994; MABROUK et al. 1995; SHEPARD et al. 1996). The dsRNA binding region has been mapped to three basic motifs between residues 221 and 305 (MILLER and SAMUEL 1992). Interestingly, core particles derived from *tsG453* consistently produce less mRNA than do cores derived from any of the other reovirus *ts* clones in *in vitro* transcriptase assays (CROSS and FIELDS 1972; COOMBS 1996). Because the affected σ3 protein is not present in the transcrip-tionally active core particle, this mutant may contain additional, as yet unknown, lesions. Additional group G mutants have been recently recovered as spontaneous mutants from high-passage virus stocks (AHMED et al. 1980a) or rescued during reversion analyses (AHMED et al. 1980a; COOMBS et al. 1994). The assignment of one of these newer clones (*tsG28.22*) to group G was confirmed by reassortant mapping (Coombs et al. 1994); however, little has been reported concerning these latter mutants.

4.9 Group H

Mutants in group H have only been recovered as spontaneous mutations arising either from high-passage stocks (Ahmed et al. 1980a) or during the rescue of pseudorevertants (Ramig and Fields 1979; Ahmed et al. 1980a; Coombs et al. 1994). Recombination assays with the then available groups A–G showed that these clones were able to recombine with all tested members of groups A–G, suggesting a new group, which was designated H. Reassortant mapping of the original prototypic mutant (*tsH26/8*) indicated that the defect resided in the M1 gene (Ramig et al. 1983). The M1 gene encodes minor core protein μ2 (McCrae and Joklik 1978; Mustoe et al. 1978b). Virions contain about 12 copies of this 736-amino acid (80-kDa) protein. The precise location and function(s) of this minor core protein remain unknown. Reassortant mapping experiments suggest that μ2 plays a role in determining the severity of CPE in cultured cells (Moody and Joklik 1989) and the level of virus growth in cardiac cells (Matoba et al. 1991) and in endothelial cells (Matoba et al. 1993). The protein also is involved in myocarditis (Sherry and Fields 1989), in organ-specific virulence in severe combined immunodeficient (SCID) mice (Haller et al. 1995), and in *in vitro* transcription of ssRNA (Yin et al. 1996). Thus, functional and stoichiometric considerations similar to those described above for the λ3 protein (see Sect. 4.5), suggest that μ2 may be located near the core vertices (see Fig. 2). The sequence of the M1 gene has been determined (Wiener et al. 1989a; Zou and Brown 1992b). However, the protein shares no significant homology with other proteins in databases; thus the sequence provides few clues as to the function or functions of the protein.

Little additional work has been reported for the *tsH26/8* clone. While analyzing revertants of *tsC447* (described above), we isolated eight additional spontaneous mutants (Coombs et al. 1994; see also Table 2). These mutants represented T1L × *tsC447* reassortants, and studies of one of them (*tsH11.2*) indicated that the clone's T1L-derived M1 gene contained a *ts* lesion (Coombs 1996). *tsH11.2* has an EOP value of about 10^{-4} at 39 °C. When grown at the nonpermissive temperature, the mutant fails to assemble identifiable particles and is defective in the production of progeny dsRNA, producing about 0.1% of the amount made at the permissive temperature (Coombs 1996); thus it represents another RNA$^-$ mutant. However, *tsH11.2* appears to differ from the other RNA$^-$ mutants of groups C, D, and E. As indicated in earlier sections, mutants in groups C, D, and E synthesize significantly reduced amounts of ssRNA as well. Viral ssRNA synthesis appears to be unaffected in restrictively incubated *tsH11.2*-infected cells. Kinetic studies of viral protein synthesis show that protein synthesis from the primary RNA transcripts is normal; however, because of the lack of progeny dsRNA, there are few secondary ssRNA transcripts and thus virtually no late protein synthesis (Coombs 1996).

Sequence analysis of the mutated M1 gene showed two alterations: a U to C transition at nucleotide position 1209 and a C to A transition at nucleotide position 1254 (Fig. 3a). Each of these alterations led to predicted amino acid changes: a methionine to threonine substitution at amino acid 399 and a change from proline to histidine at amino acid 414 (Fig. 3b). The first alteration is predicted to induce minor changes in secondary structure, whereas the second is predicted to result in significant changes in secondary structure, with conversion of a region of β-strand between amino acids Pro-408 and Leu-413 to an α-helix motif (COOMBS 1996) (Fig. 3c). It is not known which of the two alterations is responsible for the *ts* phenotype (or whether both are responsible).

Although advances have been made that allow the viral genetic information to be artificially added to susceptible cells to produce infectious progeny (RONER et al. 1990), engineering specific mutations into the genome remains elusive (JOKLIK and RONER 1996). Thus site-specific mutagenesis is currently not a practical means for determining the identities of the amino acids responsible for reovirus *ts* phenotypes. The identity of the individual amino acid substitutions responsible for the mutant phenotype in some *ts* mutants has been determined, either by extensive reversion analyses (e.g., *tsC447*, see Sect. 4.4) or because the implicated gene contains only a single alteration (e.g., *tsE320*, see Sect. 4.6). Neither approach is currently applicable to molecular analysis of the one or more lesions in *tsH11.2*. The defect or defects in *tsH11.2* can be complemented by growth of the mutant virus in cells that constitutively express wild-type μ2, even when such infected cells are incubated at the nonpermissive temperature (ZOU and BROWN 1996). This may provide a means to determine the identity of the amino acid or acids responsible for the mutant phenotype. Preliminary experiments with engineered L cells that express μ2 protein that contains both mutations (Fig. 4, "Double") indicate that, like normal L929 cells that do not express μ2 (L929), such cells do not efficiently complement the growth of *tsH11.2* at the nonpermissive temperature, whereas cells that express μ2 protein that contains either single amino acid correction (clones M399T and P414H) partially complement the defect in *tsH11.2* replication (S. ZOU, E.G. BROWN, and K.M. COOMBS, unpublished). The complementation by expressed μ2 is specific for defective μ2; no cell type supported the replication of *tsC447* at the nonpermissive temperature (Fig. 4). These results suggest that both mutant amino acid substitutions in the *tsH11.2* μ2 protein are important for expression of the *ts*

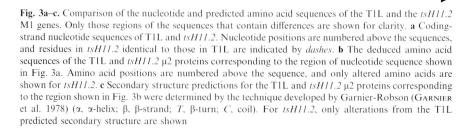

Fig. 3a–c. Comparison of the nucleotide and predicted amino acid sequences of the T1L and the *tsH11.2* M1 genes. Only those regions of the sequences that contain differences are shown for clarity. **a** Coding-strand nucleotide sequences of T1L and *tsH11.2*. Nucleotide positions are numbered above the sequences, and residues in *tsH11.2* identical to those in T1L are indicated by *dashes*. **b** The deduced amino acid sequences of the T1L and *tsH11.2* μ2 proteins corresponding to the region of nucleotide sequence shown in Fig. 3a. Amino acid positions are numbered above the sequence, and only altered amino acids are shown for *tsH11.2*. **c** Secondary structure predictions for the T1L and *tsH11.2* μ2 proteins corresponding to the region shown in Fig. 3b were determined by the technique developed by Garnier-Robson (GARNIER et al. 1978) (α, α-helix; β, β-strand; *T*, β-turn; *C*, coil). For *tsH11.2*, only alterations from the T1L predicted secondary structure are shown

a

```
        1190                                                                                          1273
T1L     AGG CAU ACA AUC GAU GUC AUG CCU GAU AUA UAU GAC UUC GUU AAA CCC AUU GGC GCU GUG CUG CCU AAG GGA UCA UUU AAA UCA
tsH11.2 --- --- --- --- --- --- -C- --- --- --- --- --- --- --- --- --- --- --- --- --- --- -A- --- --- --- --- --- ---
```

b

```
        399                                                          414     420
T1L     Arg His Thr Ile Asp Val Met Pro Asp Ile Tyr Asp Phe Val Lys Pro Ile Gly Ala Val Leu Pro Lys Gly Ser Phe Lys Ser
tsH11.2                     Thr                                                                    His
```

c

```
T1L     T T β β β β β β β β β β β β β β T T C α β β
tsH11.2   β             T                     α α α α α α α α
```

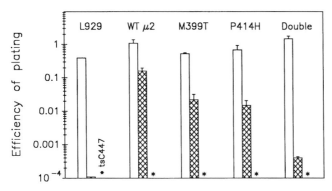

Fig. 4. Complementation of *tsH11.2* by L cells that constitutively express a variety of wild-type and mutant μ2 proteins. Efficiency of plating (EOP) values of wild-type T1L (*white bars*), *tsC447* (*), and *tsH11.2* (*hatched bars*) were determined by plating serial dilutions of the virus clones on cells that express no μ2 (L929), wild-type μ2 (WT μ2), μ2 that contains the *ts* lesion only at the first position (M399T), μ2 that contains the *ts* lesion only at the second position (P414H), or μ2 that contains both *ts* lesions (Double). Cells were incubated at 32 °C and 39 °C and EOP values determined as described in the legend to Fig. 1B. *Asterisks*, the EOP values for *tsC447* were less than 2×10^{-5} on all cell types

phenotype. Additional work to better understand this mutant and the normal role or roles of this unique protein are clearly required.

4.10 Group I

A single group I mutant (*tsI138*) has been described. This mutant was rescued from extragenically suppressed pseudorevertants (RAMIG and FIELDS 1979) and was mapped to the L3 gene (RAMIG et al. 1983), which encodes the major core protein λ1 (MCCRAE and JOKLIK 1978). Virions contain about 120 copies of this 1233-amino acid (approximately 140-kDa) protein, which is a major component of the core capsid (Fig. 2) and which is required for its assembly (XU et al. 1993). Portions of this protein are located on the exterior of the core particle (WHITE and ZWEERINK 1976). The protein has a zinc finger motif (BARTLETT and JOKLIK 1988) and a separate region that binds dsRNA (LEMAY and DANIS 1994). Other functions of the protein are not known, although recent work suggests that the protein has ATPase activity, which probably is involved in transcriptional events (NOBLE and NIBERT 1997). The *tsI138* mutant has proven difficult to manipulate; although it has an EOP value of about 10^{-3} at 39 °C, it generally grows to low titer, even at permissive temperature. The poor growth of this mutant also may explain the generally low recombination values obtained when this clone is crossed with mutants in other groups (for examples, see AHMED et al. 1980b; COOMBS 1996; Table 2). Preliminary experiments with this mutant suggest that, under nonpermissive conditions, the synthesis of protein is only marginally reduced (G. LEMAY, personal communication), but that the synthesis of dsRNA is significantly reduced (P.R. HAZELTON and K.M. COOMBS, unpublished; G. LEMAY, personal communication); thus *tsI138* may represent another RNA⁻ mutant. Electron microscopy examination of restrictively infected cells does not reveal intact viral particles (P.R. HAZELTON and K.M. COOMBS, unpublished; G. LEMAY, personal communication).

4.11 Group J

The last group of reovirus *ts* mutants were also rescued from extragenically suppressed pseudorevertants (RAMIG and FIELDS 1979). The prototype mutant *tsJ128* was mapped to the S1 gene (RAMIG et al. 1983). The S1 gene is bicistronic and encodes the approximately 50-kDa cell attachment protein σ1 (McCRAE and JOKLIK 1978; MUSTOE et al. 1978b; WEINER et al. 1978; LEE et al. 1981b; NIBERT et al. 1990). It also encodes nonstructural protein σ1s, a 14-kDa protein with unknown functions (for a review, see NIBERT et al. 1996b), but which has recently been shown to elicit a cytotoxic T lymphocyte response (HOFFMAN et al. 1996). The *tsJ128* mutant also has been difficult to manipulate; its EOP value is indistinguishable from that of T3D. This poor EOP also may explain the generally low recombination values obtained when this clone is crossed with mutants in other groups (see Table 2). Recent work with this mutant has not been reported.

5 Elucidation of Reovirus Replication by Using Temperature-Sensitive Mutants

As indicated earlier, the ability to use conditionally lethal virus mutants would be expected to shed insight onto processes affected by the gene or genes of interest. In this way, as described above, it has been possible to assign functions to some of the genes. In addition, the altered phenotypes of the mutants under nonpermissive conditions can be reconciled with known or suspected functions of the proteins, whether determined as described above or by other genetic, biochemical, or biophysical methods. Significant detail has emerged with regard to the probable replicative pathways of reovirus (for examples, see SHATKIN and KOZAK 1983; ZARBL and MILLWARD 1983; JOKLIK and RONER 1996). The initial step in this process (shown diagrammatically in Fig. 5a) after the virus has recognized and entered a susceptible host cell is removal of the outer capsid proteins (uncoating) to produce the transcriptionally active core particle. If either entry or uncoating is prevented (e.g., by a defective μ1 protein, such as is found in the restrictively assembled *tsA279* mutant), then all subsequent steps should be arrested. This blockade in entry/uncoating could result in the accumulation of mutant virions in vesicles, as observed (HAZELTON and COOMBS 1995). After successful entry and uncoating, kinetic studies indicate that core particles initially transcribe only the L1, M3, S3, and S4 genes to produce the corresponding capped l1, m3, s3, and s4 mRNA and corresponding λ3, μNS, σNS, and σ3 proteins (WATANABE et al. 1968; NONOYAMA et al. 1974; LAU et al. 1975). These proteins then appear to promote the

a

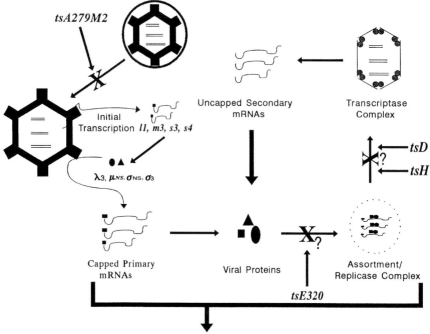

b

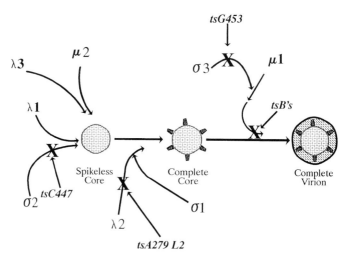

Fig. 5a, b. Model for **a** reovirus replication and **b** capsid shell assembly. For details, see text. The presumptive blocks in replication and assembly mediated by various *ts* mutants are indicated by *X* at several steps in the pathways. The *thickness* of the straight pathway *arrows* represents the amount of material generated (or used) during that step. In **a**, only some of the dsRNA segments (*double lines* inside core particle and complexes), ssRNA (*single wavy lines*), and viral proteins (*solid squares, circles,* and *triangles*) are shown for clarity. The *small solid squares* on the left end of the ssRNA (*left*) represent the mRNA cap structure transcription of all ten genes to produce all ten capped mRNA and the full complement of structural and nonstructural proteins. Although the slight reduction in amounts of ssRNA and the significant reduction in amounts of progeny dsRNA produced by mutants in groups D (L1 gene, λ3 protein) and E (S3 gene, σNS protein) could be explained by the roles these proteins play at this early step in replication, the observation that all ten species of ssRNA are produced by these mutants at the nonpermissive temperature (CROSS and FIELDS 1972; ITO and JOKLIK 1972a) suggests that the roles affected by the lesions are later. One likely site might be when newly synthesized σNS normally binds to the newly synthesized ssRNA (HUISMANS and JOKLIK 1976; GOMATOS et al. 1981), a process that is thought to facilitate genomic assortment, condensation, and second-strand synthesis (ANT-CZAK and JOKLIK 1992; JOKLIK and RONER 1996). Thus, a defective σNS protein (*tsE320*) could have little effect upon ssRNA synthesis, but could significantly reduce dsRNA synthesis, as has been observed (CROSS and FIELDS 1972; ITO and JOKLIK 1972a). Additional kinetic studies suggest that the dramatic increase seen in protein synthesis late in infection results from later transcription from the newly generated replicase–transcriptase complexes (WATANABE et al. 1968; SHATKIN and KOZAK 1983; COOMBS 1996). Because transcription appears to take place in core and core-like subviral particles (CHANG and ZWEERINK 1971; JOKLIK 1972; SHATKIN and LAFIANDRA 1972; SILVERSTEIN et al. 1972; DRAYNA and FIELDS 1982a; YIN et al. 1996), defects in any of the proteins that constitute the core, such as the major structural ones (*tsC*, σ2; *tsI*, λ1), or the enzymatic proteins (*tsD*, λ3; *tsH*, μ2), might seriously impair subsequent dsRNA synthesis, thereby accounting for the RNA⁻ phenotypes of each of these mutants. Finally, the σ3 protein appears to play an important role in the switch from early cap-dependent translation to late cap-independent translation (SKUP and MILLWARD 1980; ZARBL et al. 1980; SKUP et al. 1981; LEMIEUX et al. 1987); thus, potentially explaining the observation that *tsG453* produces less ssRNA, dsRNA, and protein than do the other RNA⁺ mutants (CROSS and FIELDS 1972; SHING and COOMBS 1996).

Analyses of the above, as well as the other, reovirus *ts* mutants also has made it possible to construct probable assembly pathways of the protein capsids by examining the morphologies of particles produced by respective mutants at the nonpermissive temperature (Fig. 5b). For example, the observation that *tsC447* fails to assemble core particles at the nonpermissive temperature (FIELDS et al. 1971, COOMBS et al. 1994; Fig. 5b, lower left) confirms the importance of the σ2 protein in core capsid assembly (XU et al. 1993; COOMBS et al. 1994). However, the ability of *tsC447* to assemble empty outer capsid structures at the restrictive temperature (FIELDS et al. 1971; MATSUHISA and JOKLIK 1974; COOMBS et al. 1994) indicates that prior assembly of the core is not required for condensation of the outer capsid and that the outer capsid proteins contain sufficient information for their own assembly. The association of core spike protein λ2 with these mutant outer capsids (MATSUHISA and JOKLIK 1974) also indicates extensive protein–protein interactions between the core spikes and the outer capsid. Mutations in core spike protein λ2 appear to be responsible for the inability of most group B mutants' assembly to proceed past a core particle (FIELDS et al. 1971; MORGAN and

ZWEERINK 1974), supporting the conclusion of core spike–outer capsid protein interactions. In addition, other mutations in λ2 (e.g., the *tsA279* L2 mutation) prevent attachment of the λ2 spike onto the core capsid shell (P.R. HAZELTON and K.M. COOMBS, manuscript in preparation). Finally, mutations in outer capsid protein σ3 appear to prevent association of this protein with the other outer capsid protein, µ1 (Fig. 5b, upper right), suggesting that a σ3–µ1 association is a prerequisite for outer capsid condensation (SHING and COOMBS 1996). Thus, numerous mutants exist which affect multiple steps in the replicative and morphogenetic pathways.

The reovirus *ts* mutants also have proven useful in various studies to elucidate the requirements for the initiation of progeny dsRNA synthesis. For example, in one detailed and elegant study, ANTCZAK and JOKLIK (1992) sought to understand the mechanism of genome segment assortment. This process, which gives rise to reassortants, is believed to involve the single-stranded mRNA before it is converted into progeny dsRNA. The investigators used RNA⁻ mutants in groups C, D, and E at the nonpermissive temperature in order to "trap" the ssRNA and to allow them to study the association of specific viral proteins with the viral mRNA. These studies indicated that 10–30 copies of nonstructural protein µNS associates initially with each of the ten mRNAs to generate ssRNA–protein complexes. These complexes then recruit σNS as well as outer capsid protein σ3. These studies also indicated that protein λ2 is added to the complexes at about the time the ssRNA is converted to dsRNA (for greater detail, see ANTCZAK and JOKLIK 1992; JOKLIK and RONER 1996).

6 Comparison of Reovirus *ts* Mutants with *ts* Mutants of Other *Reoviridae*

ts mutants have been isolated and characterized for other members of the *Reoviridae* family. These include the avian reoviruses (HAFFER 1984), rotaviruses (for recent reviews, see GOMBOLD and RAMIG 1994; ESTES 1996), and the orbiviruses (for a recent review, see ROY 1996). *ts* mutants that correspond to ten of the predicted 11 simian rotavirus SA11 recombination groups have been described (GREENBERG et al. 1981; RAMIG 1982, 1983b). Many of these have been mapped to specific genome segments (GOMBOLD et al. 1985; GOMBOLD and RAMIG 1987), and their abilities to synthesize viral RNA and protein and to assemble particles, along with other biological properties, have been determined (RAMIG and PETRIE 1984; CHEN et al. 1990; RAMIG and GOMBOLD 1991; VASQUEZ et al. 1993; GOMBOLD and RAMIG 1994; MANSELL et al. 1994; SANDINO et al. 1994; MUNOZ et al. 1995; RIOS et al. 1995). *ts* mutants also have been isolated from rhesus rotavirus (RRV) (KALICA et al. 1983) and from bovine rotavirus strain UK (GREENBERG et al. 1981; FAULKNER-VALLE et al. 1982, 1983); however, these latter sets comprise fewer of the genome segments. *ts* mutants also have been isolated from a number of different orbiviruses, including bluetongue virus, serotype 10 (BTV-10) (Shipham and de la Rey 1976), Broadhaven virus (Moss and Nuttall 1986), and Wallal and Mudjinbarry viruses (Gorman et al. 1978). The BTV-10 *ts* mutants were assigned to six of the expected ten groups (Shipham 1979), and the abilities of some of the mutants to produce RNA and protein were determined (Shipham 1979; Shipham and De la Rey 1979). The Broadhaven *ts* mutants have also been partially characterized (Moss et al. 1987, 1988; Nuttall et al. 1989).

Not surprisingly, a number of similarities exist between the types and phenotypes of the various *ts* mutants in these groups of dsRNA viruses. For example, complementation between different *ts* groups of rotavirus also is inefficient (Ramig 1982), possibly because of the abilities of the mutants to interfere with the growth of wild-type virus (Ramig 1983a, b). In addition, studies of these mutants have revealed that reassortment is an early event in the replication cycles of reovirus (Fields 1971) and rotavirus (Ramig 1983a), as well as in influenza virus (MacKenzie 1970). For these reasons, genetic recombination to generate ts^+ progeny from crosses between two different *ts* mutants appears to generate "all-or-none" values for all of the *Reoviridae* (Fields and Joklik 1969; Shipham and de la Rey 1976; Ramig 1982). Similar genetic interactions have been observed in other viruses with segmented genomes, such as the *Orthomyxoviridae* (for a recent review, see Murphy and Webster 1996) and the *Bunyaviridae* (for a review, see Bishop and Shope 1979). For a more detailed discussion of the analyses of rotavirus *ts* mutants, see Gombold and Ramig (1994).

Dissimilarities also may exist between the reoviruses and other members of the *Reoviridae*. Rotavirus *ts* mutants were used as a rapid screen to examine the phenomenon of superinfection exclusion (Ramig 1990). These studies suggested that the rotaviruses are not subject to superinfection exclusion, in contrast to the orbiviruses, in which superinfection exclusion does appear to restrict the ability of the virus to undergo reassortment (el-Hussein et al. 1989; Ramig et al. 1989). Although the use of *ts* mutants of the mammalian reoviruses to address the issue of superinfection exclusion has not been reported, other analyses suggest that superinfection exclusion does not play a role in the replication of the avian reoviruses (Ni and Kemp 1990).

7 Future Directions

The ability to increase the *ts* phenotype by slightly increasing the restrictive temperature of incubation has made it possible to more reliably examine a greater panel of *ts* mutants. This was necessary for the few virus clones (i.e., *tsA279*, *tsE320*, *tsF556*, and possibly *tsJ128*) for which EOP values at the classical non-permissive temperature of 39 °C were not significantly different from wild-type EOP values, resulting in "leaky" phenotypes which were difficult to interpret. Historically, these mutants had not been well studied, and some represent proteins for which the functions remain essentially unknown. Thus, study of these mutants under conditions where the mutant phenotype is clearly expressed should provide better insight into the functions of the affected proteins. In addition, recent advances in the ability to rapidly sequence and clone these genes should also provide a more detailed molecular understanding of the roles played by these proteins and how these roles are affected by the lesions. Continued refinement of the infectious RNA system (RONER et al. 1990), combined with a better understanding of the minimal requirements for inserting foreign or modified gene sequences into the virus (ZOU and BROWN 1992a; RONER et al. 1995), should lead to a useable system that allows *in vitro* replication, transcription, and assembly, as is being developed in the related rotaviruses (PATTON 1986; D CHEN et al. 1994; ZENG et al. 1996). Finally, higher-resolution structural determinations of the various viral and subviral particles and of the various protein components, possibly by combining information from electron cryomicroscopy and X-ray crystallography, as has been done with bluetongue virus (GRIMES et al. 1995), rhinoviruses (SMITH et al. 1996), some plant viruses (SPEIR et al. 1995), and enveloped alphaviruses (CHENG et al. 1995), should provide us with a structural framework into which information can be fitted. The combination of these improved biophysical, genetic, and biochemical techniques with the established methods of reassortant analyses and the power of *ts* mutant study should provide us with a more detailed molecular understanding of the roles played by these proteins and allow us to better delineate the replicative cycle and assembly pathways of this virus.

Note added in proof In the course of preparing this chapter, we examined the phenomenon of reovirus superinfection exclusion, which, as explained at the end of Sect. 6, had been examined among some other members of the family *Reoviridae*, but had not been studied in reovirus. We used a variety of reovirus *ts* mutants as a rapid screen for superinfection exclusion, essentially as described for rotavirus *ts* mutants (RAMIG 1990). Our results indicated that superinfecting reoviruses are not excluded from mouse L929 cells (the cell line normally used for reovirus propagation and study); nor are they excluded from monkey kidney MA104 cells; nor from human astrocytes, enterocytes, and macrophages (KEIRSTEAD and COOMBS, submitted).

References

Ahmed R, Fields BN (1981) Reassortment of genome segments between reovirus defective interfering particles and infectious virus: construction of temperature-sensitive and attenuated viruses by rescue of mutations from DI particles. Virology 111:351–363

Ahmed R, Chakraborty PR, Fields BN (1980a) Genetic variation during lytic virus infection: high passage stocks of wild-type reovirus contain temperature-sensitive mutants. J Virol 34:285–287

Ahmed R, Chakraborty PR, Graham AF, Ramig RF, Fields BN (1980b) Genetic variation during persistent reovirus infection: presence of extragenically suppressed temperature-sensitive lesions in wild-type virus isolated from persistently infected cells. J Virol 34:383–389

Antczak JB, Joklik WK (1992) Reovirus genome segment assortment into progeny genomes studied by the use of monoclonal antibodies directed against reovirus proteins. Virology 187:760–776

Astell C, Silverstein SC, Levin DH, Acs G (1972) Regulation of the reovirus RNA transcriptase by a viral capsomere protein. Virology 48:648–654

Atwater JA, Munemitsu SM, Samuel CE (1986) Biosynthesis of reovirus-specified polypeptides. Molecular cDNA cloning and nucleotide sequence of the reovirus serotype 1 Lang strain s4 mRNA which encodes the major capsid surface polypeptide sigma 3. Biochem Biophys Res Commun 136:183–192

Bartlett JA, Joklik WK (1988) The sequence of the reovirus serotype 3 L3 genome segment which encodes the major core protein lambda 1. Virology 167:31–37

Bartlett NM, Gillies SC, Bullivant S, Bellamy AR (1974) Electron microscope study of reovirus reaction cores. J Virol 14:315–326

Bergmann JE (1989) Using temperature-sensitive mutants of VSV to study membrane protein biogenesis. Methods Cell Biol 32:85–110

Bishop DHL, Shope RE (1979) Bunyaviridae. In: Fraenkel-Conrat H, Wagner RR (eds) Comprehensive virology, vol 14. Plenum, New York, pp 1–156

Black LW, Showe MK, Steven AC (1994) Morphogenesis of the T4 head. In: Karam JD, Drake JW, Kreuzer KN, Mosig G, Hall D, Eiserling FA, Black LW, Kutter E, Spicer E, Carlson K, Miller ES (eds) Molecular biology of bacteriophage T4. American Society for Microbiology, Washington DC, pp 218–258

Carleton M, Brown DT (1996) Events in the endoplasmic reticulum abrogate the temperature sensitivity of Sindbis virus mutant ts23. J Virol 70:952–959

Cashdollar LW (1994) Characterization and structural localization of the reovirus lambda 3 protein. Res Virol 145:277–285

Chakraborty PR, Ahmed R, Fields BN (1979) Genetics of reovirus: the relationship of interference to complementation and reassortment of temperature-sensitive mutants at non-permissive temperatures. Virology 94:119–127

Chang C-T, Zweerink HJ (1971) Fate of parental reovirus in infected cell. Virology 46:544–555

Chen D, Gombold JL, Ramig RF (1990) Intracellular RNA synthesis directed by temperature-sensitive mutants of simian rotavirus SA11. Virology 178:143–151

Chen D, Zeng CQ, Wentz MJ, Gorziglia M, Estes MK, Ramig RF (1994) Template-dependent, in vitro replication of rotavirus RNA. J Virol 68:7030–7039

Chen H, Ramachandra M, Padmanabhan R (1994) Biochemical characterization of a temperature-sensitive adenovirus DNA polymerase. Virology 205:364–370

Cheng RH, Kuhn RJ, Olson NH, Rossmann MG, Choi HK, Smith TJ, Baker TS (1995) Nucleocapsid and glycoprotein organization in an enveloped virus. Cell 80:621–630

Cleveland DR, Zarbl H, Millward S (1986) Reovirus guanylyltransferase is L2 gene product lambda 2. J Virol 60:307–311

Compton SR, Nelsen B, Kirkegaard K (1990) Temperature-sensitive poliovirus mutant fails to cleave VP0 and accumulates provirions. J Virol 64:4067–4075

Coombs KM (1996) Identification and characterization of a double-stranded RNA-reovirus temperature-sensitive mutant defective in minor core protein μ2. J Virol 70:4237–4245

Coombs KM, Mak SC, Petrycky-Cox LD (1994) Studies of the major reovirus core protein sigma 2: reversion of the assembly-defective mutant tsC447 is an intragenic process and involves back mutation of Asp-383 to Asn. J Virol 68:177–186

Creighton TE (1990) Protein folding. Biochem J 270:1–16

Cross RK, Fields BN (1972) Temperature-sensitive mutants of reovirus 3: studies on the synthesis of viral RNA. Virology 50:799–809

Danis C, Garzon S, Lemay G (1992) Further characterization of the ts453 mutant of mammalian orthoreovirus serotype 3 and nucleotide sequence of the mutated S4 gene. Virology 190:494–498

Denzler KL, Jacobs BL (1994) Site-directed mutagenic analysis of reovirus sigma 3 protein binding to dsRNA. Virology 204:190–199

Dermody TS, Schiff LA, Nibert ML, Coombs KM, Fields BN (1991) The S2 gene nucleotide sequences of prototype strains of the three reovirus serotypes: characterization of reovirus core protein sigma 2. J Virol 65:5721–5731

Drayna D, Fields BN (1982a) Activation and characterization of the reovirus transcriptase: genetic analysis. J Virol 41:110–118

Drayna D, Fields BN (1982b) Biochemical studies on the mechanism of chemical and physical inactivation of reovirus. J Gen Virol 63:161–170

Drayna D, Fields BN (1982c) Genetic studies on the mechanism of chemical and physical inactivation of reovirus. J Gen Virol 63:149–159

Dryden KA, Wang G, Yeager M, Nibert ML, Coombs KM, Furlong DB, Fields BN, Baker TS (1993) Early steps in reovirus infection are associated with dramatic changes in supramolecular structure and protein conformation: analysis of virions and subviral particles by cryoelectron microscopy and image reconstruction. J Cell Biol 122:1023–1041

el-Hussein A, Ramig RF, Holbrook FR, Beaty BJ (1989) Asynchronous mixed infection of Culicoides variipennis with bluetongue virus serotypes 10 and 17. J Gen Virol 70:3355–3362

Ericsson M, Cudmore S, Shuman S, Condit RC, Griffiths G, Krijnse Locker J (1995) Characterization of ts16, a temperature-sensitive mutant of vaccinia virus. J Virol 69:7072–7086

Estes MK (1996) Rotaviruses and their replication. In: Fields BN, Knipe DM, Howley PM, Chanock RM, Melnick JL, Monath TP, Roizman B, Strauss SE (eds) Virology. Lippincott-Raven, Philadelphia, pp 1625–1655

Fane B, King J (1991) Intragenic suppressors of folding defects in the P22 tailspike protein. Genetics 127:263–277

Faulkner-Valle GP, Clayton AV, McCrae MA (1982) Molecular biology of Rotaviruses. III. Isolation and characterization of temperature-sensitive mutants of bovine rotavirus. J Virol 42:669–677

Faulkner-Valle GP, Lewis J, Pedley S, McCrae MA (1983) Isolation and characterization of ts mutants of bovine rotavirus. In: Compans RW, Bishop DHL (eds) Double-stranded RNA viruses. Elsevier, New York, pp 303–312

Fausnaugh J, Shatkin AJ (1990) Active site localization in a viral mRNA capping enzyme. J Biol Chem 265:7669–7672

Fields BN (1971) Temperature-sensitive mutants of reovirus type 3: features of genetic recombination. Virology 46:142–148

Fields BN (1973) Genetic reassortment with reovirus mutants. In: Fox CF, Robinson WS (eds) Virus research. Proceedings of the 2nd ICN-UCLA symposium on molecular biology. Academic, New York, pp 461–479

Fields BN, Joklik WK (1969) Isolation and preliminary genetic and biochemical characterization of temperature-sensitive mutants of reovirus. Virology 37:335–342

Fields BN, Raine CS, Baum SG (1971) Temperature-sensitive mutants of reovirus type 3: defects in viral maturation as studied by immunofluorescence and electron microscopy. Virology 43:569–578

Fields BN, Laskov R, Scharff MD (1972) Temperature-sensitive mutants of reovirus type 3: studies on the synthesis of viral peptides. Virology 50:209–215

Furuichi Y, Shatkin AJ (1977) 5'-Termini of reovirus mRNA: ability of viral cores to form caps post-transcriptionally. Virology 77:566–578

Furuichi Y, Morgan M, Muthukrishnan S, Shatkin AJ (1975) Reovirus messenger RNA contains a methylated blocked 5'-terminal structure M7G(5')ppp(5')GmpCp-. Proc Natl Acad Sci USA 72:362–366

Garnier J, Osguthorpe DJ, Robson B (1978) Analysis of the accuracy and implications of simple method for predicting the secondary structure of globular proteins. J Mol Biol 120:97–120

Giantini M, Shatkin AJ (1989) Stimulation of chloramphenicol acetyltransferase mRNA translation by reovirus capsid polypeptide sigma 3 in cotransfected COS cells. J Virol 63:2415–2421

Giantini M, Seliger LS, Furuichi Y, Shatkin AJ (1984) Reovirus type 3 genome segment S4: nucleotide sequence of the gene encoding a major virion surface protein. J Virol 52:984–987

Gomatos PJ, Prakash O, Stamatos NM (1981) Small reovirus particles composed solely of sigma NS with specificity for binding different nucleic acids. J Virol 39:115–124

Gombold JL, Ramig RF (1987) Assignment of simian rotavirus SA11 temperature-sensitive mutant groups A, C, F, and G to genome segments. Virology 161:463–473

Gombold JL, Ramig RF (1994) Genetics of the rotaviruses. Curr Top Microbiol Immunol 185:129–177

Gombold JL, Estes MK, Ramig RF (1985) Assignment of simian rotavirus SA11 temperature-sensitive mutant groups B and E to genome segments. Virology 143:309–320

Goodenough DA, Goliger JA, Paul DL (1996) Connexins, connexons, and intracellular communication. Annu Rev Biochem 65:475–502

Gordon CL, King J (1994) Genetic properties of temperature-sensitive folding mutants of the coat protein of phage P22. Genetics 136:427–438

Gorman BM, Taylor J, Brown K, Young PR (1978) The isolation of recombinants between related orbiviruses. J Gen Virol 41:333–342

Greenberg HB, Kalica AR, Wyatt RW, Jones RW, Kapikian AZ, Chanock RM (1981) Rescue of noncultivable human rotavirus by gene reassortment during mixed infection with ts mutants of the cultivable bovine rotavirus. Proc Natl Acad Sci USA 78:420–424

Grimes J, Basak AK, Roy P, Stuart D (1995) The crystal structure of bluetongue virus VP7. Nature (Lond) 373:167–170

Haffer K (1984) In vitro and in vivo studies with an avian reovirus derived from a temperature-sensitive mutant clone. Avian Dis 28:669–676

Haller BL, Barkon ML, Vogler GP, Virgin HW 4th (1995) Genetic mapping of reovirus virulence and organ tropism in severe combined immunodeficient mice: organ-specific virulence genes. J Virol 69:357–364

Hazelton PR, Coombs KM (1995) The reovirus mutant tsA279 has temperature-sensitive lesions in the M2 and L2 genes: The M2 gene is associated with decreased viral protein production and blockade in transmembrane transport. Virology 207:46–58

Hoffman LM, Hogan KT, Cashdollar LW (1996) The reovirus nonstructural protein sigma1NS is recognized by murine cytotoxic T lymphocytes. J Virol 70:8160–8164

Huismans H, Joklik WK (1976) Reovirus-coded polypeptides in infected cells: isolation of two native monomeric polypeptides with high affinity for single-stranded and double-stranded RNA, respectively. Virology 70:411–424

Ikegami N, Gomatos PJ (1968) Temperature-sensitive conditional-lethal mutants of reovirus 3. I. Isolation and characterization. Virology 36:447–458

Imani F, Jacobs BL (1988) Inhibitory activity for the interferon-induced protein kinase is associated with the reovirus serotype 1 sigma 3 protein. Proc Natl Acad Sci USA 85:7887–7891

Ito Y, Joklik WK (1972a) Temperature-sensitive mutants of reovirus. I. Patterns of gene expression by mutants of groups C, D, and E. Virology 50:189–201

Ito Y, Joklik WK (1972b) Temperature-sensitive mutants of reovirus. II. Anomalous electrophoretic migration behavior of certain hybrid RNA molecules composed of mutant plus strands and wild-type minus strands. Virology 50:202–208

Jayasuriya AK, Nibert ML, Fields BN (1988) Complete nucleotide sequence of the M2 gene segment of reovirus type 3 dearing and analysis of its protein product mu 1. Virology 163:591–602

Joklik WK (1972) Studies on the effect of chymotrypsin on reovirions. Virology 49:700–701

Joklik WK, Roner MR (1995) What reassorts when reovirus genome segments reassort? J Biol Chem 270:4181–4184

Joklik WK, Roner MR (1996) Molecular recognition in the assembly of the segmented reovirus genome. Prog Nucleic Acid Res Mol Biol 53:249–281

Kalica AR, Flores J, Greenberg HB (1983) Identification of the rotaviral gene that codes for hemagglutination and protease-enhanced plaque formation. Virology 125:194–205

Keirstead ND, Coombs KM Absence of superinfection exclusion during asynchronous infections of mouse, moneky, and human cell lines (submitted)

Lau RY, Van Alstyne D, Berckmans R, Graham AF (1975) Synthesis of reovirus-specific polypeptides in cells pretreated with cyclohexamide. J Virol 16:470–478

Lee PWK, Hayes EC, Joklik WK (1981a) Characterization of antireovirus immunoglobulins secreted by cloned hybridoma cell lines. Virology 108:134–146

Lee PWK, Hayes EC, Joklik WK (1981b) Protein 1 is the reovirus cell attachment protein. Virology 108:156–163

Lemay G, Danis C (1994) Reovirus lambda 1 protein: affinity for double-stranded nucleic acids by a small amino-terminal region of the protein independent from the zinc finger motif. J Gen Virol 75:3261–3266

Lemieux R, Lemay G, Millward S (1987) The viral protein sigma 3 participates in translation of late viral messenger RNA in reovirus-infected L cells. J Virol 61:2472–2479

Lucia-Jandris P, Hooper JW, Fields BN (1993) Reovirus M2 gene is associated with chromium release from mouse L cells. J Virol 67:5339–5345

Luftig RB, Kilham SS, Hay AJ, Zweerink HJ, Joklik WK (1972) An ultrastructure study of virions and cores of reovirus type 3. Virology 48:170–181

Mabrouk T, Lemay G (1994) Mutations in a CCHC zinc-binding motif of the reovirus sigma 3 protein decrease its intracellular stability. J Virol 68:5287–5290

Mabrouk T, Danis C, Lemay G (1995) Two basic motifs of reovirus sigma 3 protein are involved in double-stranded RNA binding. Biochem Cell Biol 73:137–145

MacKenzie JS (1970) Isolation of temperature-sensitive mutants and the construction of a preliminary map for influenza virus. J Gen Virol 6:63–75

Mansell EA, Ramig RF, Patton JT (1994) Temperature-sensitive lesions in the capsid proteins of the rotavirus mutants tsF and tsG that affect virion assembly. Virology 204:69–81

Mao ZX, Joklik WK (1991) Isolation and enzymatic characterization of protein lambda 2, the reovirus guanylyltransferase. Virology 185:377–386

LIVERPOOL
JOHN MOORES UNIVERSITY
AVRIL ROBARTS LRC
TEL. 0151 231 4022

Matoba Y, Sherry B, Fields BN, Smith TW (1991) Identification of the viral genes responsible for growth of strains of reovirus in cultured mouse heart cells. J Clin Invest 87:1628–1633

Matoba Y, Colucci WS, Fields BN, Smith TW (1993) The reovirus M1 gene determines the relative capacity of growth of reovirus in cultured bovine aortic endothelial cells. J Clin Invest 92:2883–2888

Matsuhisa T, Joklik WK (1974) Temperature-sensitive mutants of reovirus. V. Studies on the nature of the temperature-sensitive lesion of the group C mutant ts447. Virology 60:380–389

McCoy KL (1990) Contribution of endosomal acidification to antigen processing. Semin Immunol 2:239–246

McCrae MA, Joklik WK (1978) The nature of the polypeptide encoded by each of the ten double-stranded RNA segments of reovirus type 3. Virology 89:578–593

McPhillips TH, Ramig RF (1984) Extragenic suppression of temperature-sensitive phenotype in reovirus: mapping suppressor mutations. Virology 135:428–439

Metcalf P, Cyrklaff M, Adrian M (1991) The three-dimensional structure of reovirus obtained by cryo-electron microscopy. EMBO J 10:3129–3136

Miller JE, Samuel CE (1992) Proteolytic cleavage of the reovirus sigma 3 protein results in enhanced double-stranded RNA-binding activity: identification of a repeated basic amino acid motif within the C-terminal binding region. J Virol 66:5347–5356

Millns AK, Carpenter MS, Delange AM (1994) The vaccinia virus-encoded uracil DNA glycosylase has an essential role in viral DNA replication. Virology 198:504–513

Mitraki A, King J (1992) Amino acid substitutions influencing intracellular protein folding pathways. FEBS Lett 307:20–25

Moody MD, Joklik WK (1989) The function of reovirus proteins during the reovirus multiplication cycle: analysis using monoreassortants. Virology 173:437–446

Morgan EM, Zweerink HJ (1974) Reovirus morphogenesis. Corelike particles in cells infected at 39° with wild-type reovirus and temperature sensitive mutants of groups B and G. Virology 59:556–565

Morozov SY (1989) A possible relationship of reovirus putative RNA polymerase to polymerases of positive-strand RNA viruses. Nucleic Acids Res 17:5394

Moss SR, Nuttall PA (1986) Isolation and characterization of temperature sensitive mutants of Broadhaven virus, a Kemerovo group orbivirus (family, reoviridae). Virus Res 4:331–336

Moss SR, Ayres CM, Nuttall PA (1987) Assignment of the genome segment coding for the neutralizing epitope(s) of orbiviruses in the Great Island subgroup (Kemerovo serogroup). Virology 157:137–144

Moss SR, Ayres CM, Nuttall PA (1988) The Great Island subgroup of tick-borne orbiviruses represents a single gene pool. J Gen Virol 69:2721–2727

Munoz M, Rios M, Spencer E (1995) Characteristics of single- and double-stranded RNA synthesis by a rotavirus SA-11 mutant thermosensitive in the RNA polymerase gene. Intervirology 38:256–263

Murphy BR, Webster RG (1996) Orthomyxoviruses. In: Fields BN, Knipe DM, Howley PM, Chanock RM, Melnick JL, Monath TP, Roizman B, Straus SE (eds) Fields virology, vol 1. Lippincott-Raven, Philadelphia, pp 1397–1445

Murphy BR, Prince GA, Collins PL, Van-Wyke-Coelingh K, Olmsted RA, Spriggs MK, Parrott RH, Kim HW, Brandt CD, Chanock RM (1988) Current approaches to the development of vaccines effective against parainfluenza and respiratory syncytial viruses. Virus Res 11:1–15

Mustoe TA, Ramig RF, Sharpe AH, Fields BN (1978a) A genetic map of reovirus. III. Assignment of the double-stranded RNA mutant groups A, B, and G to genome segments. Virology 85:545–556

Mustoe TA, Ramig RF, Sharpe AH, Fields BN (1978b) Genetics of reovirus: identification of the dsRNA segments encoding the polypeptides of the mu and sigma size classes. Virology 89:594–604

Nagy PD, Dzianott A, Ahlquist P, Bujarski JJ (1995) Mutations in the helicase-like domain of protein 1a alter the sites of RNA-RNA recombination in brome mosaic virus. J Virol 69:2547–2556

Ni YW, Kemp MC (1990) Selection of genome segments following coinfection of chicken fibroblasts with avian reoviruses. Virology 177:625–633

Nibert ML, Fields BN (1992) A carboxy-terminal fragment of protein mu 1/mu 1C is present in infectious subvirion particles of mammalian reoviruses and is proposed to have a role in penetration. J Virol 66:6408–6418

Nibert ML, Dermody TS, Fields BN (1990) Structure of the reovirus cell-attachment protein: a model for the domain organization of sigma 1. J Virol 64:2976–2989

Nibert ML, Margraf RL, Coombs KM (1996a) Non-random segregation of parental alleles in reovirus reassortants. J Virol 70:7295–7300

Nibert ML, Schiff LA, Fields BN (1996b) Reoviruses and their replication. In: Fields BN, Knipe DM, Howley PM, Chanock RM, Melnick JL, Monath TP, Roizman B, Straus SE (eds) Fields virology. Lippincott-Raven, Philadelphia, pp 1557–1596

Noble S, Nibert ML (1997) Characterization of an ATPase activity in reovirus cores and its genetic association with core-shell protein lambda1. J Virol 71:2182–2191

Nonoyama M, Millward S, Graham AF (1974) Control of transcription of the reovirus genome. Nucleic Acids Res 1:373–385

Nuttall PA, Moss SR, Jones LD, Carey D (1989) Identification of the major genetic determinant for neurovirulence of tick-borne orbiviruses. Virology 172:428–434

Ogasawara N, Moriya S, Yoshikawa H (1991) Initiation of chromosome replication: structure and function of oriC and DnaA protein in eubacteria. Res Microbiol 142:851–859

Patton JT (1986) Synthesis of simian rotavirus SA11 double-stranded RNA in a cell-free system. Virus Res 6:217–233

Prevelige PE Jr, King J (1993) Assembly of bacteriophage P22: a model for ds-DNA virus assembly. Prog Med Virol 40:206–221

Pringle CR (1996) Temperature-sensitive mutant vaccines. In: Robinson A, Graham GH, Wiblin CN (eds) Methods in molecular medicine: vaccine protocols. Humana, Totowa, pp 17–32

Ramig R, Fields BN (1983) Genetics of reovirus. In: Joklik W (ed) The reoviridae. Plenum, New York, pp 197–228

Ramig RF (1982) Isolation and genetic characterization of temperature-sensitive mutants of simian rotavirus SA11. Virology 120:93–105

Ramig RF (1983a) Factors that affect genetic interaction during mixed infection with temperature-sensitive mutants of simian rotavirus SA11. Virology 127:91–99

Ramig RF (1983b) Isolation and genetic characterization of temperature-sensitive mutants that define five additional recombination groups in simian rotavirus SA11. Virology 130:464–473

Ramig RF (1990) Superinfecting rotaviruses are not excluded from genetic interactions during asynchronous mixed infections in vitro. Virology 176:308–310

Ramig RF, Fields BN (1979) Revertants of temperature-sensitive mutants of reovirus: evidence for frequent extragenic suppression. Virology 92:155–167

Ramig RF, Gombold JL (1991) Rotavirus temperature-sensitive mutants are genetically stable and participate in reassortment during mixed infection of mice. Virology 182:468–474

Ramig RF, Petrie BL (1984) Characterization of temperature-sensitive mutants of simian rotavirus SA11: protein synthesis and morphogenesis. J Virol 49:665–673

Ramig RF, Ward RL (1991) Genomic segment reassortment in rotaviruses and other reoviridae. Adv Virus Res 39:163–207

Ramig RF, Mustoe TA, Sharpe AH, Fields BN (1978) A genetic map of reovirus. II. Assignment of the double-stranded RNA-negative mutant groups C, D, and E genome segments. Virology 85:531–544

Ramig RF, Ahmed R, Fields BN (1983) A genetic map of reovirus: assignment of the newly defined mutant groups H, I, and J to genome segments. Virology 125:299–313

Ramig RF, Garrison C, Chen D, Bell-Robinson D (1989) Analysis of reassortment and superinfection during mixed infection of Vero cells with bluetongue virus serotypes 10 and 17. J Gen Virol 70:2595–2603

Rios M, Munoz M, Spencer E (1995) Antiviral activity of phosphonoformate on rotavirus transcription and replication. Antiviral Res 27:71–83

Rixon FJ, Addison C, McLauchlan J (1992) Assembly of enveloped tegument structures (L particles) can occur independently of virion maturation in herpes simplex virus type 1-infected cells. J Gen Virol 73:277–284

Roner MR, Sutphin LA, Joklik WK (1990) Reovirus RNA is infectious. Virology 179:845–852

Roner MR, Lin PN, Nepluev I, Kong LJ, Joklik WK (1995) Identification of signals required for the insertion of heterologous genome segments into the reovirus genome. Proc Natl Acad Sci USA 92:12362–12366

Roy P (1996) Orbiviruses and their replication. In: Fields BN, Knipe DM, Howley PM, Chanock RM, Melnick JL, Monath TP, Roizman B, Strauss SE (eds) Virology. Lippincott-Raven, Philadelphia, pp 1709–1734

Rozinov MN, Fields BN (1996) Interference of reovirus strains occurs between the stages of uncoating and dsRNA accumulation. J Gen Virol 77:1425–1429

Sandino AM, Fernandez J, Pizarro J, Vasquez M, Spencer E (1994) Structure of rotavirus particle: interaction of the inner capsid protein VP6 with the core polypeptide VP3. Biol Res 27:39–48

Schiff LA, Nibert ML, Co MS, Brown EG, Fields BN (1988) Distinct binding sites for zinc and double-stranded RNA in the reovirus outer capsid protein sigma 3. Mol Cell Biol 8:273–283

Schuerch AR, Matsuhisa I, Joklik WK (1974) Temperature-sensitive mutants of reovirus. VI. Mutants ts447 and ts556 particles lack one or two L genome RNA segments. Intervirology 3:36–46

Schwartzberg PL, Roth MJ, Tanese N, Goff SP (1993) Analysis of a temperature-sensitive mutation affecting the integration protein of Moloney murine leukemia virus. Virology 192:673–678

Seliger LS, Giantini M, Shatkin AJ (1992) Translational effects and sequence comparisons of the three serotypes of the reovirus S4 gene. Virology 187:202–210

Sharpe AH, Fields BN (1982) Reovirus inhibition of cellular RNA and protein synthesis: role of the S4 gene. Virology 122:381–391

Sharpe AH, Ramig RF, Mustoe TA, Fields BN (1978) A genetic map of reovirus. I. Correlation of genome RNAs between serotypes 1, 2 and 3. Virology 84:63–74

Shatkin AJ (1974) Methylated messenger RNA synthesis in vitro by purified reovirus. Proc Natl Acad Sci USA 71:3204–3207

Shatkin AJ, Kozak M (1983) Biochemical aspects of reovirus transcription and translation. In: Joklik WK (ed) The reoviridae. Plenum, New York, pp 79–106

Shatkin AJ, LaFiandra AJ (1972) Transcription by infectious subviral particles of reovirus. J Virol 10:698–706

Shatkin AJ, Sipe JD, Loh PC (1968) Separation of 10 reovirus genome segments by polyacrylamide gel electrophoresis. J Virol 2:986–991

Shepard DA, Ehnstrom JG, Skinner PJ, Schiff LA (1996) Mutations in the zinc-binding motif of the reovirus capsid protein σ3 eliminate its ability to associate with capsid protein μ1. J Virol 70:2065–2068

Sherry B, Fields BN (1989) The reovirus M1 gene, encoding a viral core protein, is associated with the myocarditic phenotype of a reovirus variant. J Virol 63:4850–4856

Shikova E, Lin YC, Saha K, Brooks BR, Wong PK (1993) Correlation of specific virus-astrocyte interactions and cytopathic effects induced by ts1, a neurovirulent mutant of Moloney murine leukemia virus. J Virol 67:1137–1147

Shing M, Coombs KM (1996) Assembly of the reovirus outer capsid requires μ1/σ3 interactions which are prevented by misfolded σ3 protein in reovirus temperature-sensitive mutant tsG453. Virus Res 46:19–29

Shipham S (1979) Further characterization of the ts mutant F207 of bluetongue virus. Onderstepoort J Vet Res 46:207–210

Shipham SO, de la Rey M (1976) The isolation and preliminary genetic classification of temperature-sensitive mutants of bluetongue virus. Onderstepoort J Vet Res 43:189–192

Shipham SO, de la Rey M (1979) Temperature-sensitive mutants of bluetongue virus: genetic and physiological characterization. Onderstepoort J Vet Res 46:87

Silverstein SC, Astell C, Levin DH, Schonberg M, Acs G (1972) The mechanisms of reovirus uncoating and gene activation in vivo. Virology 47:797–806

Simpson RW, Hirst GK (1968) Temperature-sensitive mutants of influenza A virus: isolation of mutants and preliminary observations on genetic recombination and complementation. Virology 35:41–49

Skup D, Millward S (1980) Reovirus-induced modification of cap dependent translation in infected L cells. Proc Natl Acad Sci USA 77:152–156

Skup D, Zarbl H, Millward S (1981) Regulation of translation in L-cells infected with reovirus. J Mol Biol 151:35–55

Smith TJ, Chase ES, Schmidt TJ, Olson NH, Baker TS (1996) Neutralizing antibody to human rhinovirus 14 penetrates the receptor-binding canyon. Nature 383:350–354

Spandidos DA, Graham AF (1975a) Complementation of defective reovirus by ts mutants. J Virol 15:954–963

Spandidos DA, Graham AF (1975b) Complementation between temperature-sensitive and deletion mutants of reovirus. J Virol 16:1444–1453

Speir JA, Munshi S, Wang G, Baker TS, Johnson JE (1995) Structures of the native and swollen forms of cowpea chlorotic mottle virus determined by X-ray crystallography and cryo-electron microscopy. Structure 3:63–78

Starnes MC, Joklik WK (1993) Reovirus protein lambda 3 is a poly(C)-dependent poly(G) polymerase. Virology 193:356–366

Sturzenbecker LJ, Nibert M, Furlong D, Fields BN (1987) Intracellular digestion of reovirus particles requires a low pH and is an essential step in the viral infectious cycle. J Virol 61:2351–2361

Tillotson L, Shatkin AJ (1992) Reovirus polypeptide sigma 3 and N-terminal myristoylation of polypeptide mu 1 are required for site-specific cleavage to mu 1C in transfected cells. J Virol 66:2180–2186

Tosteson MT, Nibert ML, Fields BN (1993) Ion channels induced in lipid bilayers by subvirion particles of the nonenveloped mammalian reoviruses. Proc Natl Acad Sci USA 90:10549–10552

Tyler KL, Fields BN (1996) Reoviruses. In: Fields BN, Knipe DM, Howley PM, Chanock RM, Melnick JL, Monath TP, Roizman B, Straus SE (eds) Fields virology. Lippincott-Raven, Philadelphia, pp 1597–1623

Vasquez M, Sandino AM, Pizarro JM, Fernandez J, Valenzuela S, Spencer E (1993) Function of rotavirus VP3 polypeptide in viral morphogenesis. J Gen Virol 74:937–941

Watanabe Y, Millward S, Graham AF (1968) Regulation of transcription of the reovirus genome. J Mol Biol 36:107–123

Weiner HL, Ramig RF, Mustoe TA, Fields BN (1978) Identification of the gene coding for the hemagglutinin of reovirus. Virology 86:581–584

White CK, Zweerink HJ (1976) Studies on the structure of reovirus cores: selective removal of polypeptide λ2. Virology 70:171–180

Wiener JR, Joklik WK (1987) Comparison of the reovirus serotype 1, 2, and 3 S3 genome segments encoding the nonstructural protein sigma NS. Virology 161:332–339

Wiener JR, Joklik WK (1988) Evolution of reovirus genes: a comparison of serotype 1, 2, and 3 M2 genome segments, which encode the major structural capsid protein mu 1C. Virology 163:603–613

Wiener JR, Bartlett JA, Joklik WK (1989a) The sequences of reovirus serotype 3 genome segments M1 and M3 encoding the minor protein mu 2 and the major nonstructural protein mu NS, respectively. Virology 169:293–304

Wiener JR, McLaughlin T, Joklik WK (1989b) The sequences of the S2 genome segments of reovirus serotype 3 and of the dsRNA-negative mutant ts447. Virology 170:340–341

Wiskerchen M, Muesing MA (1995) Human immunodeficiency virus type 1 integrase: effects of mutations on viral ability to integrate, direct viral gene expression from unintegrated viral DNA templates, and sustain viral propagation in primary cells. J Virol 69:376–386

Xu P, Miller SE, Joklik WK (1993) Generation of reovirus core-like particles in cells infected with hybrid vaccinia viruses that express genome segments L1, L2, L3, and S2. Virology 197:726–731

Yin P, Cheang M, Coombs KM (1996) The M1 gene is associated with differences in the temperature optimum of the transcriptase activity in reovirus core particles. J Virol 70:1223–1227

Zarbl H, Millward S (1983) The reovirus multiplication cycle. In: Joklik WK (ed) The reoviridae. Plenum, New York, pp 107–196

Zarbl H, Skup D, Millward S (1980) Reovirus progeny subviral particles synthesize uncapped mRNA. J Virol 34:497–505

Zeng CQ-Y, Wentz MJ, Cohen J, Estes MK, Ramig RF (1996) Characterization and replicase activity of double-layered and single-layered rotavirus-like particles expressed from baculovirus recombinants. J Virol 70:2736–2742

Zou S, Brown EG (1992a) Identification of sequence elements containing signals for replication and encapsidation of the reovirus M1 genome segment. Virology 186:377–388

Zou S, Brown EG (1992b) Nucleotide sequence comparison of the M1 genome segment of reovirus type 1 Lang and type 3 Dearing. Virus Res 22:159–164

Zou S, Brown EG (1996) Stable expression of the reovirus mu2 protein in mouse L cells complements the growth of a reovirus ts mutant with a defect in its M1 gene. Virology 217:42–48

Zweerink HJ, Joklik WK (1970) Studies on the intracellular synthesis of reovirus-specified proteins. Virology 41:501–518

Suppression and Reversion of Mutant Phenotype in Reovirus

R.F. RAMIG

Division of Molecular Virology, Baylor College of Medicine, One Baylor Plaza, Houston, TX 77030,
USA

1 Introduction

The concept of genetic suppression of phenotypes has existed since the original description of suppression of eye color in *Drosophila* (STURDEVANT 1920). Subsequently, genetic suppression was recognized and characterized in a variety of prokaryotic and eukaryotic organisms (GORINI and BECKWITH 1966; HARTMAN and ROTH 1973) and was demonstrated to occur through a variety of mechanisms (GORINI 1970; HARTMAN and ROTH 1973). Of particular use for genetic analysis of prokaryotic viruses were host-dependent suppressor-sensitive mutations, which were conditionally lethal depending on the host bacterial strain used as an indicator for viral growth (EDGAR 1966). The first documentation of genetic suppression in an animal virus was in reovirus (RAMIG et al. 1977). Suppression was subsequently documented in a number of animal viruses, including papovavirus (SHORTLE et al. 1979), picornavirus (KING et al. 1980), vaccinia virus (McFADDEN et al. 1980), influenza virus (TOLPIN et al. 1981), herpesvirus (HALL and ALMEY 1982), and adenovirus (KRUIJER et al. 1983), demonstrating the general applicability of the concept to animal virus genetics.

1.1 Definitions

Genetic suppression is defined as the reversal of a mutant phenotype by a second mutation at a site distinct from the site of the original mutation. Since reversal of mutant phenotype is also characteristic of reversion, the two processes must be carefully distinguished. Several possibilities exist for mechanisms by which revertant phenotype can occur:

1. Reversion can occur by precise back-mutation, restoring the wild-type nucleotide sequence of the gene and amino acid sequence of the gene product (*true reversion*).
2. Reversion can occur by mutation at the site of the original mutation so that an amino acid distinct from that contained in the mutant or the wild type is encoded at the site of the original mutation (*pseudoreversion*). The effect of the new mutation is functional activity of the encoded protein that is wild type or near enough to wild type to score as wild type.
3. Revertant phenotype can result from mutation at a second site in the same gene as the original mutation (*intragenic suppression*). For example, a second frame-shift mutation can correct the effects of an original frame-shift mutation. The effect of intragenic suppressor mutations is to functionally repair the lesion induced by the original mutation.
4. Revertant phenotype can result from mutation at a second site in a gene different from the gene containing the original mutation (*extragenic suppression*). In this case, the effect of the suppressor mutation is to bypass the functional defect resulting from the original mutation. Mechanistically, extragenic suppression is often mediated through protein–protein or protein–nucleic acid interactions.

The critical differentiating factor between true revertants and pseudorevertants on the one hand, and intragenically suppressed revertants and extragenically suppressed revertants on the other hand, is the absence of the original mutation in the former and the continued presence of the original mutation in the latter.

In this review the following terminology will be used to carefully differentiate the various types of virus with revertant phenotype. *Revertant* will refer to any virus that has gained the wild-type phenotype, regardless of mechanism. Revertants that have gained revertant phenotype by true reversion will be referred to as *true revertants* and those that have gained revertant phenotype by pseudoreversion will be called *pseudorevertants*. True revertants and pseudorevertants no longer contain the original mutation. Revertants that have gained revertant phenotype by intragenic suppression will be called *intragenically suppressed revertants* and those that have gained revertant phenotype by extragenic suppression will be called *extragenically suppressed revertants*. Intragenically suppressed revertants and extragenically suppressed revertants still contain the original mutation.

1.2 Operational Definition of Genetic Suppression

True revertants and pseudorevertants are distinguished from intragenically and extragenically suppressed revertants by *backcross* of the revertant to the parental wild-type virus. In the backcross, recombination can separate the suppressor mutation present in an intragenically or extragenically suppressed revertant from the original mutation, so that the original mutation can be recovered among the progeny of the backcross. In contrast, in true revertants or pseudorevertants, the original mutation no longer exists so it cannot be recovered from among the backcross progeny.

The operational definition for suppression requires that the virus undergo recombination at a reasonable frequency during mixed infection. This requirement is met for viruses with DNA genomes where recombination can be detected between adjacent nucleotides (Benzer 1957), but becomes problematic for viruses with RNA genomes. Although many viruses with nonsegmented RNA genomes have recently been demonstrated to undergo intramolecular recombination (Lai 1992), the genomes are relatively small and the rates of recombination are correspondingly low. The low rates of intramolecular recombination in the nonsegmented RNA genome viruses make screening of progeny from backcrosses for the rare recombinationally separated original mutant a daunting process. Intramolecular recombination has not been demonstrated for viruses with segmented RNA genomes, although they do reassort the individual genome segments at high frequency (Ramig and Fields 1983; Ramig and Ward 1991). These viruses present a special case. Upon backcross of a revertant to wild type, the genome segments reassort at high frequency. If the original mutation and the suppressor mutation lie on a single genome segment of the revertant, they cannot be separated by reassortment; however, if they lie on different genome segments of the revertant, reassortment can separate them. In this case, some of the progeny of the backcross will contain the original mutation alone so that its phenotype can be expressed and detected. Thus, in segmented genome viruses, suppression can be demonstrated by classical genetic techniques only if the suppressor mutation lies on a genome segment different from the segment bearing the original mutation (e.g., only extragenic suppression can be demonstrated by backcross). In segmented genome viruses, backcrosses of revertant to wild type which do not yield the original mutation provide no information about the nature of the intragenic event that led to the revertant phenotype; it could be true reversion, pseudoreversion, or intragenic suppression. These intragenic reversions of unspecified mechanism will be referred to as *intragenic reversion events*.

1.3 Use of Conditionally Lethal Mutations for Studies of Suppression

From the discussion above, it is clear that studies of reversion and suppression rely on the availability of easily and rapidly detectable viral phenotypes. Thus the majority of the work on suppression of animal virus mutations, and all of the work on suppression of reovirus mutations, has relied on the use of conditionally lethal mutations of the temperature-sensitive (*ts*) type. The studies of suppression of reovirus mutants described below used only mutants of the *ts* type, as they are easily scored and are easily distinguished from revertants with wild-type (ts^{+}) phenotype.

2 Features of Reovirus Biology Relevant to Studies of Suppression

Several features of reovirus biology are relevant to studies of suppression of reovirus *ts* mutations. These features impinge upon the limitations of classical genetic analyses to demonstrate suppression and upon models for mechanisms of suppression.

2.1 Genome Structure and Unusual Linkage Relationships

The genome of reoviruses consists of ten segments of double-stranded RNA. These segments reassort freely and at high frequency in mixed infection (FIELDS and JOKLIK 1969; FIELDS 1971; RAMIG and WARD 1991). The result of reassortment is that progeny genomes can contain mixtures of segments originating from either of the parental viruses (SHARPE et al. 1978). Because of the high frequency of reassortment, backcrosses of extragenically suppressed revertants to wild type separate the original and suppressor mutations at high frequency, facilitating the isolation of the original mutation from among the progeny of the backcross (RAMIG et al. 1977). In contrast, intragenically suppressed revertants contain an original mutation and a suppressor mutation that are absolutely linked because intramolecular recombination does not occur in reovirus. Thus backcrosses of intragenically suppressed revertants to wild type cannot separate the original mutation from the suppressor, and no progeny with the original mutation alone can be isolated from the progeny.

2.2 Protein–Protein Interactions During Reovirus Infection

Many of the models for mechanisms of suppression in animal viruses involve protein–protein or protein–nucleic acid interactions (RAMIG and FIELDS 1979; SHORTLE et al. 1979; RAMIG 1980). Thus an understanding of these interactions facilitates an understanding of suppression mechanisms. Relatively little is known about the interactions between reovirus proteins and nucleic acids in the infected cell. The reader is referred to other chapters in this book for information on these interactions (see the chapters by COOMBS, LEE and GILMORE, SCHIFF, and JACOBS and LANGLAND, this volume). More is known of the interactions of the reovirus proteins in the virion. In the virion, the proteins are arranged in a double-layered capsid, and the proteins of each capsid layer obviously interact in the formation of each capsid layer. It is likely that the proteins present in each capsid layer also interact with proteins in the adjacent capsid layer. The reader is referred to the chapter by NIBERT (this volume) for detailed information on the arrangement of the structural proteins. Since several of the reovirus proteins have been documented to be multifunctional, and most of them are likely to be so, it is likely that functional interactions of specific proteins remain to be elucidated. Studies of suppression may provide a tool to reveal such interactions (see Sect. 3.4).

2.3 Temperature-Sensitive Mutants

Conditionally lethal mutations of the *ts* type have been isolated from reovirus serotype 3 and characterized genetically (FIELDS and JOKLIK 1969; CROSS and FIELDS 1972; RAMIG and FIELDS 1979). Genetic analysis demonstrated that the mutants fell into ten reassortment groups (groups A–J), indicating that at least one *ts* mutation within the mutant collection resides on each of the ten genome segments (FIELDS and JOKLIK 1969; CROSS and FIELDS 1972; RAMIG and FIELDS 1979). The prototype mutant from each recombination group was physically mapped to a specific genome segment, confirming the genetic grouping of the mutants (MUSTOE et al. 1978a; RAMIG et al. 1978; RAMIG et al. 1983).

   ts mutants are defined by their ability to replicate to wild-type or near wild-type levels at permissive temperature and their inability to replicate to significant levels at nonpermissive temperature. For reovirus *ts* mutants, the permissive temperature is 31 °C and the nonpermissive temperature is 39 °C. In practical terms, *ts* mutants replicate to detectable levels at 39 °C so that, in some cases, there is no clear distinction between *ts* and ts^+ virus. To quantitate the differential between growth at 39 °C and 31 °C, the efficiency of plating (EOP) is generally used. EOP is the ratio of titer at 39 °C to titer at 31 °C; an EOP greater than 0.1 is considered indicative of ts^+ phenotype, and an EOP less than 0.1 is considered indicative of *ts* phenotype. An EOP of 0.1 is generally used to differentiate *ts* from ts^+ viruses in the studies of reovirus suppression, although in the one case in which a large number of plaques were examined, a bimodial distribution of EOP was evident, with the distinction between *ts* and ts^+ falling around an EOP of 0.01 (Fig. 1; RAMIG et al. 1977). As a result of this finding, plaques with an EOP greater than 0.1 were considered ts^+ and those with an EOP less than 0.01 were considered *ts* on the

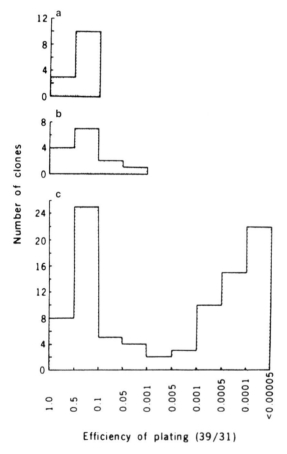

Fig. 1a–c. Two populations of progeny virus in the backcross between the reovirus revertant *RtsA(201)-101* and wild type. Shown is the distribution of temperature phenotype (efficiency of plating) versus the number of clones with that phenotype for self-crosses of **a** wild type and **b** *RtsA(201)-101* or **c** the backcross between wild type and *RtsA(201)-101*. (From RAMIG et al. 1977, with permission)

basis of a single determination. Plaques with an EOP between 0.1 and 0.01 were retitered for confirmation of the temperature phenotype.

3 Suppression of Temperature-Sensitive Phenotype

Genetic suppression in animal viruses was first suggested during studies of reversion of reovirus *ts* mutants (Cross and Fields 1976b). Subsequent to definitive demonstration of suppression in revertants of reovirus *ts* mutants (Ramig et al. 1977), the phenomenon was extensively studied and the genetics of suppression of *ts* phenotype in reovirus remains probably the best understood of any of the animal viruses.

3.1 Studies of Reversion

Examination of viral proteins synthesized by the various reovirus *ts* mutant groups revealed that mutants representing four of the seven groups examined synthesized μ1 and μ2 proteins with aberrant electrophoretic migration at both 39 °C and 31 °C (Cross and Fields 1976a). Since μ1 and μ2 (μ2 is now called μ1C) are related by cleavage and are therefore encoded in the same genome segment (subsequently shown to be M2; Mustoe et al. 1978b), it seemed unlikely that aberrant migration of μ1/μ2 was related to the *ts* mutation of all four viruses representing different reassortment groups. To resolve the question of which of the *ts* mutant groups was related to the aberrant migration of μ1/μ2, three-factor crosses were performed of the type tsA$^-$/tsD$^+$/μ$^+$ crossed with tsD$^-$/tsA$^+$/μ$^-$. In this cross, all *ts*$^+$ progeny synthesized μ$^-$ polypeptides, indicating that the determinant of the μ$^-$ phenotype was linked to the *tsA(201)* mutation, i.e., it was on the same genome segment. No other mutant group appeared to be linked to the μ$^-$ phenotype (Cross and Fields 1976b). In an attempt to confirm the linkage of *tsA* to the μ$^-$ phenotype, spontaneous *ts*$^+$ revertants were picked from the parental stock of *tsA(201)*. Of eight *tsA(201)* revertants examined, two revertants coreverted the *ts* mutation and μ$^-$ marker, four revertants gained a μ^{-2} phenotype when the *ts* mutation reverted, and one revertant retained the μ$^-$ phenotype when the *ts* mutation reverted but also gained a σ3$^-$ phenotype. The change in μ polypeptide migration in seven of eight revertants examined supported the linkage of the *tsA(201)* mutation to the μ$^-$ phenotype (Cross and Fields 1976b). The remaining revertant (revertant clone 101, called *RtsA(201)-101*), in which the μ$^-$ phenotype was retained and a σ3$^-$ phenotype was gained, proved to be particularly interesting. The fact that *RtsA(201)-101* retained the μ$^-$ phenotype suggested that it still contained the parental *tsA(201)* lesion in suppressed form, since in all other cases reversion of the *ts* mutation was accompanied by an alteration in μ migration. This notion was supported by the fact that *RtsA(201)-101* contained an altered protein σ3 and the possibility that the alteration in σ3 reflected the presence of a mutation in the σ3-encoding gene. The mutation reflected by altered migration of σ3 could be a suppressor mutation that was suppressing the *tsA(201)* lesion and was responsible for the *ts*⁺ phenotype of *RtsA(201)-101*. Since the *tsA(201)* mutation and its putative suppressor mutation lay on different genome segments, suppression could be demonstrated by a classical backcross of *RtsA(201)-101* to wild type, and the *tsA(201)* lesion was shown to be found among the backcross progeny (Ramig et al. 1977).

3.2 Demonstration of *RtsA(201)-101* as an Extragenically Suppressed Pseudorevertant

RtsA(201)-101 was backcrossed to wild-type reovirus serotype 3 at 31 °C; the progeny of the backcross was plated at 31 °C, and random plaques were picked and passaged to high titer. As controls, *RtsA(201)-101*, *tsA(201)*, and wild type were plated at 31 °C, and random plaques were picked and passaged to high titer. The resulting clones from the backcross and the control clones were tested for temperature phenotype by plaque assay at 31 °C and 39 °C and the EOP was calculated. The results (Fig. 1) showed that the wild-type and *RtsA(201)-101* control clones exhibited the expected *ts*⁺ phenotype; *tsA(201)* control clones exhibited the expected *ts* phenotype (data not shown). In contrast, the 95 backcross progeny clones showed a bimodal distribution of temperature phenotypes (Fig. 1), with 42 of the clones exhibiting *ts*⁺ phenotype and 53 of the clones exhibiting *ts* phenotype (Ramig et al. 1977). This result unequivocally demonstrated that *RtsA(201)-101* contained a suppressed *ts* mutation.

To definitively prove that *RtsA(201)-101* contained a suppressed *tsA(201)* mutation, it was necessary to show that a portion of the *ts* clones rescued in the backcross contained *ts*A(201) mutations. This was necessary because in prokaryotic systems it had been shown that suppressed pseudorevertants of *ts* mutations often contained suppressor mutations in second genes with intrinsic *ts* phenotype (Jarvik and Botstein 1975). Accordingly, 29 of the *ts* backcross progeny clones were each crossed with either *tsA(201)* or *tsB(352)* to determine whether any of the *ts* clones contained *tsA(201)* mutations. All 29 of the *ts* clones from the backcross failed to yield *ts*⁺ progeny when crossed with *tsA(201)*, and all yielded *ts*⁺ progeny when crossed with *tsB(352)*. This result indicated that all 29 of the tested *ts* progeny clones from the backcross contained the *tsA(201)* mutation and that the suppressor mutation did not appear to have *ts* phenotype (Ramig et al. 1977).

Taken together, these results demonstrated by classical genetic techniques that *RtsA(201)-101* was an extragenically suppressed revertant of the *tsA(201)* mutation. However, these results did not answer a number of important questions, such as the following: (a) Was suppression of *ts* mutations in reovirus a common phenomenon, or was *RtsA(201)-101* an unusual revertant? (b) Can suppressor mutations have a *ts* phenotype, or can they only be detected by their suppressive activity? (c) What gene contained the mutation that suppressed *tsA(201)*? (d) What was the mechanism by which suppressor mutations exerted their effect?

3.3 Incidence of Extragenic Suppression of Reovirus *ts* Mutations

To determine whether extragenic suppression was a common means of gaining revertant phenotype in reovirus, a large number of revertants were isolated and tested. Twenty-eight independent, spontaneous revertants were isolated from mutants representing all seven of the reassortment groups of *ts* mutants identified at that time (RAMIG and FIELDS 1979). Analysis of progeny from each revertant backcrossed to wild-type reovirus type 3 revealed that 25 of the 28 revertants contained *ts* mutations, and in every case but one the parental *ts* mutation was isolated and identified by reassortment analysis. This result demonstrated that most revertants of reovirus *ts* mutants were extragenically suppressed revertants, because the operational definition of suppression was met in 25 of 28 revertants examined.

Reassortment analysis of the *ts* mutations isolated from the revertant/wild-type backcrosses revealed that some of the rescued *ts* mutations were not representative of the original *ts* mutant group, but represented other reassorting mutant groups (RAMIG and FIELDS 1979). While some of the *ts* mutations rescued from the backcross were in the same mutant group as the parental mutation, or in other reassortment groups, others reassorted with all seven of the mutant groups included in the analysis, indicating that they represented *ts* mutations in reassortment groups not previously identified. When the new *ts* mutants were grouped among themselves by reassortment, three reassortment groups were identified that had not been previously identified (RAMIG and FIELDS 1979). The addition of three mutant groups to the seven existing mutant groups brought the total number of *ts* mutant groups for reovirus to ten, the expected number. Subsequently, the prototype mutant of each *ts* mutant group was physically mapped to one of the ten reovirus genome segments (MUSTOE et al. 1978a; RAMIG et al. 1978, 1983).

During the studies performed to map the newly identified *ts* mutant reassortment groups to genome segments, confounding data was obtained until it was realized that these *ts* mutants easily reentered the suppressed state during mapping crosses (RAMIG et al. 1983). Clear map positions were obtained for each of the three new mutant groups when it was demonstrated that a significant number of the ts^+ mapping clones actually contained the *ts* lesion in suppressed form. However, definitive mapping required the tedious demonstration of extragenic suppression in a large number of mapping clones. One byproduct of these studies was the demonstration that suppression could occur in intertypic reassortant clones, which suggested that suppressor mutations were not serotype specific (RAMIG et al. 1983).

Whether any of the nonparental *ts* mutants rescued in the revertant/wild-type backcrosses represented extragenic suppressor mutations with *ts* phenotype has not been determined. Such a determination would require the construction of a double *ts* mutant containing both the putative *ts* suppressor mutation and the original *ts* mutation and demonstration that the double mutant had ts^+ phenotype, a tedious task.

Several other conclusions were reached from these studies (RAMIG and FIELDS 1979).

1. Extragenically suppressed revertants appeared to be very stable in their revertant phenotype. A total of 237 clones were derived from self-crosses of the 28 revertants, and all of them were ts^+. Thus the revertant phenotype was stable, and back-mutation of the suppressor mutation to wild type, while theoretically possible, did not occur at a detectable frequency.
2. The high frequency of reversion by extragenic suppression suggested a general mechanism by which RNA viruses lacking DNA intermediates in their life cycles could overcome the effects of deleterious mutations. Since these viruses cannot generate viable combinations of genetic material from nonviable genomes by intramolecular recombination, generation of extragenic suppressor mutations would provide a means of bypassing the effects of mutations that accumulated in the absence of intramolecular recombination. Extragenic suppressor mutations in segmented genome viruses could be rapidly disseminated throughout the virus population by reassortment.
3. Extragenically suppressed revertants were found for every mutant and mutant group examined. Thus there seemed to be no correlation between mutant site and the ability to become suppressed.
4. A parallel study of ts^+ virus clones isolated from an L cell line persistently infected with $tsC(447)$ revealed some of them to be extragenically suppressed revertants, indicating that suppressor mutations can be selected by means other than temperature selection (AHMED et al. 1980).
5. The finding that many of the nonparental ts mutations rescued in backcrosses of revertants to wild type represented mutations in new reassortment groups suggested that rescue of ts progeny from extragenically suppressed revertants might provide a means of selecting new mutations in systems in which some mutants existed but the map had not been saturated. Similar results were previously reported in bacteriophage genetic systems (FLOOR 1970; JARVIK and BOTSTEIN 1975), but had not been reported in animal viruses.
6. These results provided no additional insights as to the mechanism by which the suppressor mutations acted.

3.4 Mechanisms of Extragenic Suppression

Numerous mechanisms have been proposed for the action of suppressor mutations (GORINI and BECKWITH 1966; GORINI 1970; HARTMAN and ROTH 1973). *Direct, intergenic suppression* occurs when a mutation in a different gene from the original mutation causes a change that circumvents rather than repairs the original lesion. A striking example of direct intergenic suppression was demonstrated in bacteriophage P22, in which suppressors of ts mutations were mapped to genes whose products interacted with the product of the original mutation (JARVIK and BOT-STEIN 1975), or bacteriophage T4, in which suppressors restored stoichiometric relationships altered by the original mutation (FLOOR 1970). *Direct, intragenic suppression* occurs when a mutation in the same gene as the original mutation restores the function of the product of the gene containing the two mutations. Examples of direct, intragenic suppression are the restoration of reading frame by a second frame-shift mutation in a single gene (CRICK et al. 1961) or the suppressor of a deletion in the influenza virus NS1 protein which restored proper folding of the protein (TREANOR et al. 1991). *Direct, intergenic suppression* occurs when the function of a mutant product of the gene containing the original mutation is restored by an alteration in a second gene that changes the fidelity of information transmission (often called informational suppression). The classic example of such suppression is the suppression of termination codons (amber mutations) by host-encoded tRNA that insert an amino acid in response to the UGA termination codon. Finally, drug-induced informational suppression is a special case of direct, intergenic suppression in which aminoglycoside antibiotics alter the fidelity of information transmission by allowing ribosomes to accept normally unacceptable codon–anticodon pairs during translation (GORINI 1970).

Although no direct evidence exists to support any of the above mechanisms in suppression of reovirus *ts* mutations, it can be argued that direct, intragenic or intergenic mechanisms do not operate in extragenically suppressed reovirus revertants (RAMIG and FIELDS 1979). Direct, intragenic mechanisms can be discarded, because virtually all studies of suppression of *ts* mutations in reovirus showed extragenic suppressor mutations. The possibility of direct, intergenic suppression is unlikely, because (a) reversion of *ts* mutations was not generally accompanied by an increase in size of the mutant protein, as would be expected by suppression of termination mutations, (b) missense, and not nonsense, mutations generally produce *ts* phenotypes (CAMPBELL 1961), and (c) the small number of genes, and identification of gene product for each gene, makes it unlikely that reovirus encodes genes for tRNA species. This leaves indirect, intergenic mechanisms as the most likely mode of action for reovirus extragenic suppressor mutations (RAMIG and FIELDS 1979).

In viral systems, indirect, intergenic suppression has been demonstrated to occur by the suppressor mutation restoring stoichiometric relationships altered by the original mutation (FLOOR 1970) or by the suppressor mutation producing a second (suppressor) protein that is in physical contact with the parental (*ts*) protein (JARVIK and BOTSTEIN 1975). In the absence of any stoichiometric alterations of protein synthesis in any of the reovirus *ts* mutants (CROSS and FIELDS 1976a), the second of these hypotheses must be favored. Two variations of this hypothesis have been proposed (RAMIG 1980):

1. The *compensating protein interactions* model (Fig. 2) predicts that the *ts* mutation causes an alteration in the structure of the encoded protein such that the protein interacts with a second protein in a manner that is thermolabile. In the extragenically suppressed revertant, the second protein is also altered (by the

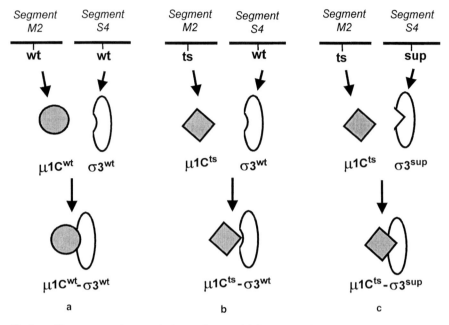

Fig. 2a–c. The compensating protein interactions model for extragenic suppression in reovirus. The model is diagrammed as it would apply to the suppressed pseudorevertant *RtsA(201)-101*. **a** In wild type, the wild-type μ1C and σ3 proteins interact to form a thermostable complex. **b** In *tsA(201)*, the interaction between *ts*-μ1C and wild-type σ3 proteins produces a thermolabile complex. **c** In the pseudorevertant *RtsA(201)-101*, the interaction of *ts*-μ1C with σ3 containing a suppressor mutation restores the thermostability of the complex

suppressor mutation) so that its interaction with the original protein is thermostable. These compensating interactions yield a complex that is thermostable, although both protein components are mutant. In reovirus, where many of the gene products are structural components of the virion, suppressor mutations would be expected in any protein with which the *ts* protein interacts, either in the virion or during morphogenesis. For example, the compensating protein interactions model would predict that extragenic suppressors of the *tsA(201)* mutation would map to any gene encoding a product that interacted with protein μ1/μ1C, the product of the M2 segment bearing the *tsA(201)* mutation. In the virion, μ1C and σ3 are the major components of the outer capsid and would be expected to interact. Thus one would expect that suppressors of *tsA(201)* might well map to the S4 segment and its encoded σ3 product. Indeed, in the original revertant clone investigated (*RtsA(201)-101*; see Sect. 3.2), suppression was suspected because the revertant had both an altered μ1C and an altered σ3 (CROSS and FIELDS 1976b; RAMIG et al. 1977). The location of the suppressor mutation in extragenically suppressed revertant *RtsA(201)-101* was subsequently demonstrated to be segment S4 (MCPHILLIPS and RAMIG 1984).

2. In the *mutator transcriptase* model, the *ts* mutation makes a protein thermolabile in its interactions with another protein. The defect in the interactions between

proteins is corrected by a compensating alteration in a second protein, as described above. However, in this case, the second, suppressor mutation is the direct result of a virion transcriptase that reads with lower fidelity than the wild-type enzyme. Since the transcription products of reovirus function as both mRNA and template for the synthesis of progeny genomes, any mutations introduced by the mutator transcriptase could be "locked in" to the genome of progeny virions. By chance, some progeny virions would contain the combination of a *ts* mutation and a suppressor mutation and would express revertant phenotype. This model is attractive in that it would account for both the high frequency of extragenically suppressed pseudorevertants among revertants of reovirus *ts* mutants and the high frequency of nonparental *ts* mutations that are isolated from suppressed pseudorevertants (see Sects. 3.3, 4.2.2).

3.5 Mapping Extragenic Suppressor Mutations

The interacting proteins model for extragenic suppression in reovirus stimulated interest in mapping extragenic suppressor mutations because, if the model were correct, mapping original mutation–suppressor mutation pairs might reveal previously unknown protein–protein interactions. Suppressor mutations were mapped by both classical genetic and biochemical methods.

3.5.1 Genetic Mapping

Suppressor mutations present in extragenically suppressed reovirus pseudorevertants were mapped (McPHILLIPS and RAMIG 1984) by a variation of the methods used to map the locations of the prototype *ts* mutations (MUSTOE et al. 1978a; RAMIG et al. 1978). Revertants of reovirus type 3 *ts* mutations, previously determined to contain extragenic suppressor mutations (RAMIG and FIELDS 1979), were backcrossed to wild-type reovirus type 1. Because the two parental viruses in the backcrosses represented different serotypes, the parental origin of each genome segment in progeny clones could be determined by electrophoresis (Fig. 3; SHARPE et al. 1978). The progeny of the backcross were screened for those that met two criteria:

1. The clone should contain and express the parental *ts* mutation (e.g., a progeny clone from a revertant derived from *tsA(201)* should express *tsA(201)* phenotype). This criterion required that the suppressor mutation in the clone had reverted or that it had reassorted so that the clone had obtained the wild-type cognate of the segment bearing the suppressor mutation.
2. Since clones in which the suppressor mutation had reverted would be uninformative for mapping, the clone should also be reassortant.

Once a collection of clones meeting these criteria had been isolated from a backcross, the parental origin of the genome segments in the clones could be analyzed to map the location of the suppressor mutation (Fig. 3).

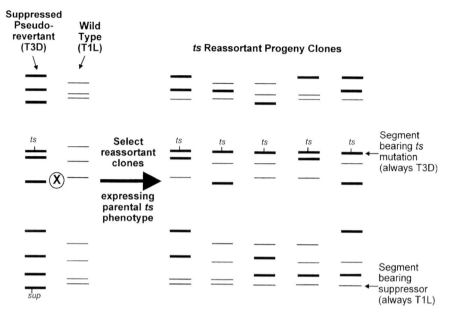

Fig. 3. Backcross to map the location of the suppressor mutation in an extragenically suppressed pseudorevertant, illustrating the case for mapping the suppressor of *RtsA(201)-101*. Shown are the genomes of the parental viruses and mapping clones as they appear following electrophoretic separation to allow the assignment of parental origin for each genome segment

Using this methodology, suppressor mutations were mapped to genome segments for four extragenically suppressed revertants and narrowed to two segments for two other revertants (McPhillips and Ramig 1984). The inability to precisely map the location of the suppressor mutation in two revertants was the result of the small number of clones isolated from these backcrosses that met the criteria for mapping. Mapping of the suppressor mutation present in extragenically suppressed pseudo-revertant *RtsA(201)-101* will be discussed in detail and the others summarized.

Twenty-five reassortants expressing the *tsA(201)* phenotype were isolated when 652 random progeny plaques from the backcross of *RtsA(201)-101* and wild-type reovirus type 1 were analyzed (McPhillips and Ramig 1984). Of the 652 randomly chosen progeny plaques, 128 expressed a *ts* phenotype in a screening assay. Electrophoresis revealed that 69 of the 128 were reassortant. Upon passage of the 69 *ts*/reassortant plaques to high titer, nine lost *ts* phenotype, leaving 60 high-titer clones that were both *ts* and reassortant. Standard reassortment grouping revealed that only 25 of the *ts*/reassortant clones met the criteria of being *ts* in the parental reassortment group and being reassortant. Thus only 3.83% (25 out of 652) of the progeny screened yielded clones possessing properties consistent with being useful for mapping the suppressor mutation.

Analysis of the parental origin of genome segments in the 25 mapping clones derived from the backcross of *RtsA(201)-101* revealed the following (McPhillips and Ramig 1984):

- All 25 clones contained genome segment M2 of reovirus type 3 origin. This result was expected and was consistent with the map position of the reovirus type 3 *tsA(201)* mutation, which had previously been established as segment M2 (Mustoe et al. 1978a).
- Twenty-four of the 25 mapping clones contained genome segment S4 derived from the wild-type reovirus type 1 parent. This result strongly suggested that the suppressor mutation present in *RtsA(201)-101* resided on genome segment S4. One clone contained segment S4 of type 3 origin, suggesting that the suppressor mutation had reverted in that clone.
- The remaining eight genome segments were derived randomly from either parental virus, indicating that none of them could contain either the *tsA(201)* mutation or the suppressor mutation.

Taken together, these data indicated that the suppressor of the *tsA(201)* mutation present in *RtsA(201)-101* mapped to genome segment S4. Since segment M2 encodes the μ1C protein and S4 encodes the σ3 protein (Mustoe et al. 1978b), it confirmed the original hypothesis that the suppressor present in *RtsA(201)-101* would map to segment S4 because the revertant contained an altered σ3 protein (see Sect. 3.1). These map locations for *tsA(201)* and the suppressor mutation are consistent with the compensating protein interactions model of extragenic suppression, because μ1C and σ3 are the major components of the reovirus outer capsid and must interact intimately in the formation of the capsid structure.

Suppressor mutations were mapped for three other revertants (McPhillips and Ramig 1984). The suppressor present in revertant *RtsA(201)-121* also mapped to segment S4, confirming results obtained with *RtsA(201)-101*. The suppressor present in revertant *RtsA(201)-122* mapped to genome segment L3. In the compensating protein interactions model of suppression, this result suggests that μ1 and/or μ1C (products of segment M2 bearing the *tsA(201)* mutation) interacts with protein λ1, which is encoded in segment L3. μ1/μ1C is not known to interact with λ1 in the virion, where the former are components of the outer capsid and the latter is a minor component of the core. However, it is possible that these proteins interact while performing functions unrelated to their structural roles in the virion. One suppressor of the *tsB(352)* mutation was mapped. The suppressor present in *RtsB(352)b* mapped to segment S1. The suppressor-containing segment S1 encodes protein σ1, and the *tsB(352)*-containing segment L2 encodes protein λ2 (McCrae and Joklik 1978; Mustoe et al. 1978a). This result also fits with the compensating protein interactions model of suppression, as close juxtaposition of σ1 and λ2 in the virion has been documented (Lee et al. 1981). Data from two other extragenically suppressed revertants allowed the position of the suppressor mutations to be narrowed to one of two segments. Too few useful mapping clones were derived from backcrosses with these revertants to allow a definitive map position to be determined (McPhillips and Ramig 1984). The suppressor mutation of revertant *RtsA(340)b* mapped to either segment L1 or segment S4. In revertant *RtsA(438)c*, the suppressor mapped to either segment L3 or S4.

Several features of the genetic mapping data (McPHILLIPS and RAMIG 1984) merit additional comment.

– Reversion of suppressor mutations had not been documented, although reversion of a suppressor was invoked to make the mapping data obtained for the suppressor in *RtsA(201)-101* consistent. One clone was obtained from backcross of another extragenically suppressed revertant that contained the parental *ts* mutation and also had all its genome segments derived from reovirus type 3. This clone provided the first proof that suppressor mutations could revert (McPHILLIPS and RAMIG 1984).
– The backcrosses of extragenically suppressed revertants to wild type in the mapping studies yielded many fewer progeny clones expressing the parental *ts* phenotype than expected from prior studies (RAMIG et al. 1977; RAMIG and FIELDS 1979). As a result, experiments which undertook to map suppressors in 18 suppressed pseudorevertants produced unambiguous map locations for suppressors in only four cases, and no clones meeting the criteria for mapping could be isolated in nine cases. One possibly important difference between the experiments to document suppression (RAMIG et al. 1977; RAMIG and FIELDS 1979) and these experiments to map suppressors was that the suppressed pseudorevertant was backcrossed to wild type of a different serotype in the mapping experiments. The intertypic nature of the crosses may have lowered the frequency of reassortment as compared to homotypic crosses, resulting in fewer clones appropriate for mapping. Alternatively, wild-type gene products from the heterotypic virus may have been able to suppress mutations. Suppression was indeed identified in progeny from heterotypic crosses subsequently performed to map new *ts* mutations (RAMIG et al. 1983; see Sect. 3.3). Suppression of a *ts* mutation by a gene derived from a heterologous wild type has also been documented in influenza virus (GHENDON et al. 1982; SCHOLTISSEK and SPRING 1982).
– The results obtained in cases in which suppressor mutations could be mapped were generally consistent with the compensating protein interactions model for suppression.

3.5.2 Biochemical Mapping

The mapping of reovirus *ts* mutations and their suppressors to specific genome segments (McPHILLIPS and RAMIG 1984) made possible the fine mapping of *ts* mutations and suppressor mutations by sequencing the appropriate genome segments from extragenically suppressed revertants (A.K.A. Jayasuriya and B.N. Fields, personal communication; JAYASURIYA 1991). In the studies described below, sequencing was used to fine map the *tsA(201)* mutation within the M2 genome segment and the suppressors of this mutation that were mapped to segment S4 in two different extragenically suppressed revertants.

Sequence analysis of segment M2 from *tsA(201)*, and comparison to the M2 sequence of reovirus type 3 wild type, established that the μ1 protein of the mutant differed from wild type by two substitutions: Ala to Val at position 305 and Pro to Ser at position 315 (JAYASURIYA 1991). The Ala to Val substitution at position 305 is a conservative change, and Val at position 305 is also seen in other wild-type M2 sequences (WIENER and JOKLIK 1988). Thus it is unlikely to confer *ts* phenotype (JAYASURIYA 1991). In contrast, the proline at position 315 is conserved across all other M2 sequences, and the Pro to Ser change at this position substitutes a hydrophobic residue for a polar residue. On this basis, it was concluded that the *ts* phenotype of *tsA(201)* resulted from the Pro to Ser mutation at position 315. Surprisingly, three algorithms used to predict protein secondary structure indicated that no structural change in μ1/μ1C was caused by the substitution at position 315 in *tsA(201)* (JAYASURIYA 1991).

Sequence analysis of the M2 and S4 segments from extragenically suppressed revertants *RtsA(201)-101* and *RtsA(201)-121* revealed the following (JAYASURIYA 1991):

1. The M2 segments of both revertants were identical in sequence to the parental *tsA(201)* mutant M2.
2. Segment S4 of extragenically suppressed revertant *RtsA(201)-101* contained a single mutation leading to an amino acid substitution when compared to the sequence of S4 from *tsA(201)*. This Glu to Ala substitution at position 217 of the σ3 protein represents the suppressor mutation present in *RtsA(201)-101*. The predicted amino acid change at position 217 of σ3 abolished a V8 protease recognition site which had previously been mapped (SCHIFF et al. 1988), and the absence of this cleavage site was confirmed. This result suggests that the domain of σ3 around position 217 is exposed on the surface of the protein. The application of three protein-folding algorithms to the σ3 sequence from *RtsA(201)-101* indicated that the Glu to Ala substitution at position 217 abolished an α-helical sequence in the region and extended a β-sheet structure. The suppressor mutation may alter σ3 conformation so that it can interact with mutant μ1C in such a way that the complex is thermostable.
3. Segment S4 of extragenically suppressed revertant *RtsA(201)-121* contained a single mutation when compared to S4 of *tsA(201)*, leading to a Gly to Ser substitution at position 47 of the σ3 protein. The Gly to Ser change at position 47 of σ3 represents the suppressor mutation present in *RtsA(201)-121*. Application of three protein-folding algorithms to the σ3 sequence from *RtsA(201)-121* indicated that the Gly to Ser change at position 47 abolished four turns in the region of the mutation. The region surrounding position 47 of σ3 is one of the most hydrophobic in the protein, suggesting that it may be involved in interactions with other proteins, including μ1C.

In summary, sequence analysis provided precise localization of the *ts* mutation in the *tsA(201)* mutant and precise locations of the suppressor mutations in two extragenically suppressed revertants derived from *tsA(201)*. Although protein-folding algorithms did not predict a conformational change in μ1/μ1C associated with the *tsA(201)* mutation, changes in conformation of σ3 were predicted for both of the suppressor mutations (A.K.A. JAYASURIYA and B.N. FIELDS, personal communication; JAYASURIYA 1991). These results are consistent with the compensating protein interactions model of suppression.

3.6 Intragenic Suppression

Backcross of the revertant *RtsA(438)b*, derived from the *tsA(438)* mutant, failed to reveal *ts* progeny, suggesting that *RtsA(438)b* had either undergone an intragenic reversion event or was an intragenically suppressed revertant (RAMIG and FIELDS 1979). To determine the mechanism by which *RtsA(438)b* gained *ts*⁺ phenotype, it was subjected to sequence analysis (A.K.A. Jayasuriya and B.N. Fields, personal communication; JAYASURIYA 1991).

Sequence analysis of segment M2 from *tsA(438)* revealed that it contained the same two mutations as M2 from *ts*A(201), and the Pro to Ser change at position 315 of μ1C was inferred to cause the *ts* phenotype (JAYASURIYA 1991). Sequence analysis of segment M2 from *RtsA(438)b* revealed a single substitution, Ser to Arg at position 652 of μ1C, in addition to the two mutations present in μ1C of *tsA(438)*. This result indicated that *RtsA(438)b* retained the original *ts* mutation (Pro to Ser at position 315) and had reverted by generating a suppressor mutation at a second site (Ser to Arg at position 652). Thus *RtsA(438)b* was confirmed as an intragenically suppressed revertant. Protein-folding algorithms predicted that the domain surrounding position 652 was located on the surface of μ1C and that the mutation at position 652 extended the length of an α-helix in the region (JAYASURIYA 1991). The α-helical region near position 652 is amphipathic, suggesting that it may promote interaction of the domain with other regions of μ1C through hydrophobic interactions. Such intramolecular interactions could result in the stabilization of a μ1C which also contains a destabilizing (*ts*) mutation.

These results demonstrate that *ts* mutants of reovirus can revert by the generation of intragenic suppressor mutations as well as by the more common generation of extragenic suppressor mutations. No further studies of intragenic suppression in reovirus have been reported.

3.7 True Reversion

The *tsC(447)* mutation was mapped to genome segment S2, which encodes the major core protein σ2 (MUSTOE et al. 1978b; RAMIG et al. 1978). *tsC(447)* fails to assemble core particles at the nonpermissive temperature (FIELDS et al. 1971). In an attempt to understand the assembly of core particles and the proteins with which σ2 interacts during morphogenesis, ten independent revertants of *tsC(447)* were isolated so that suppressor mutations could be mapped (COOMBS et al. 1994). Each of the *tsC(447)* revertant clones was able to assemble core particles in infected cells, indicating that the effect of the original mutation had indeed been bypassed or repaired. Backcrosses of the revertants to wild-type reovirus type 1 revealed that the *tsC(447)* lesion could not be identified in any of the progeny clones derived from any of the revertants. This result indicated that *tsC(447)* had reverted exclusively by intragenic reversion events and not by extragenic suppression. To determine the molecular basis of the *ts*[+] revertant phenotype, the S2 genome segment from *tsC(447)* and from each of nine revertant clones was sequenced. The S2 segment from the parental *tsC(447)* used in these experiments contained two of the three mutations (Ala to Val at position 323 and Asn to Asp at position 383 of σ2) that had been reported previously for *tsC(447)* (WIENER et al. 1989). Although some of the revertants contained two or three mutations, all of the nine revertants examined contained the back mutation Asp to Asn at position 383 of σ2. This result indicated that the Asn to Asp mutation at position 383 of σ2 was responsible for the *ts* phenotype of *tsC(447)* and that all the revertants were true revertants with back mutations (Asp to Asn) at position 383 of σ2 (COOMBS et al. 1994). The fact that the forward mutation was an A to G change at nucleotide 1166 of segment S2 and that all nine revertants contained the precise G to A back mutation at nucleotide 1166 confirmed the true revertant nature of the *tsC(447)* revertants. The region of σ2 surrounding the mutation site at residue 383 is predicted to be exposed on the surface of the protein (DERMODY et al. 1991), where it could interact with other proteins. The failure to identify interacting proteins by suppressor mapping indicated that this domain of σ2 most likely interacts with other σ2 molecules during formation of the core. The reversion of the *tsC(447)* mutation exclusively by back mutation or true reversion suggested that only Asn at position 383 was consistent with functional σ2–σ2 interactions (COOMBS et al. 1994).

4 Significance of Suppression and Reversion

Suppression and reversion of mutations in reovirus has a number of theoretical and practical implications for genetic, biochemical, and molecular biological studies of the virus. A number of these implications are discussed below.

4.1 Evolutionary Significance

The demonstration of extragenic suppression as a major pathway of reversion in reovirus *ts* mutants suggested a means by which RNA viruses could overcome the effects of deleterious mutations (RAMIG et al. 1977; RAMIG and FIELDS 1979). Since RNA viruses have either no intramolecular recombination or recombination that occurs at extremely low rates (COOPER 1968), they are unable to easily generate viable combinations of genetic material from parental genomes that are nonviable or of reduced fitness. A high frequency of mutation to an intragenically or extragenically suppressed revertant genotype would provide a means of bypassing the deleterious effects of spontaneously occurring mutations in the absence of intramolecular recombination. In reovirus, the majority of suppressed *ts* mutations were suppressed by suppressor mutations residing on different reassorting genome segments from the original mutation *ts* mutation. Since reassortment is a highly efficient means of moving genetic information on different genome segments into new combinations, rare suppressor mutations could be rapidly spread throughout the virus population. The spread of suppressor mutations through the virus population could potentially rescue viruses with normally lethal mutations and increase the fitness of viruses with nonlethal mutations.

The quasispecies nature of RNA virus populations (HOLLAND et al. 1982), including reovirus, suggests that mutations are randomly introduced into the population at a relatively high rate by the low fidelity polymerases associated with RNA viruses. This suggests that normal virus populations may contain preexisting deleterious and suppressor mutations engaged in a race to determine whether the deleterious mutation will be lost from the population before it finds its suppressor through reassortment, so that it can be retained in the population. The preexistence of suppressor mutations in virus populations may serve to reduce the size of a virus population that can pass through a genetic bottleneck without losing viability.

Although the studies of suppression in reovirus have not provided strong evidence to support the mutator polymerase model of generating suppressor mutations, if such a mutator were present in the population it would function to increase the diversity of genes within the population that could be acted upon by evolutionary forces.

4.2 Practical Implications

In this short section, a diverse selection of topics will be discussed that relate to investigators' response to the knowledge that suppression is common in reovirus revertants, to situations not described above where suppression has been documented, and to a rapid method for screening revertants for suppression.

4.2.1 Use of Revertants as Controls

Revertants of virus mutants have commonly been used as experimental controls in biochemical and genetic studies in the reovirus system. The knowledge that reovirus mutants can be suppressed should introduce a degree of caution into this practice. Since intragenically and extragenically suppressed revertants still contain the original mutation, the possibility exists that they do not represent proper control viruses. If the suppressor mutation acts by bypassing, rather than correcting, the functional defect of the original mutation, as in the compensating protein interactions model, it is possible that some of the phenotypic effects of the original mutation may still be expressed. Thus, prior to use of revertants as controls, it is important to examine the genotype of the revertant as well as its revertant phenotype.

4.2.2 Suppression of *ts* Phenotype During Persistent Infection

A serially passaged stock of reovirus *tsC(447)* was used to establish persistent infection of L cells. Within 1 month of establishment, the persistently infected cells began to shed virus with a wild type (*ts*⁺) phenotype (AHMED and GRAHAM 1977). Since the change from *ts* to *ts*⁺ phenotype during persistent infection was relatively unusual, the persistent virus was studied to determine whether it had gained *ts*⁺ phenotype by extragenic suppression (AHMED et al. 1980). Seven *ts*⁺ clones were isolated from the persistently infected cells, and each clone was backcrossed to wild-type virus. Analysis of the backcross progeny revealed that only two of seven *ts*⁺ clones (clones 9 and 26) yielded *ts* progeny, suggesting that the other five *ts*⁺ clones had gained *ts*⁺ phenotype by intragenic reversion events (true reversion, pseudo-reversion, or intragenic suppression). Reassortment analysis with the *ts* mutations rescued from clones 9 and 26 showed that none of the rescued *ts* mutations was the parental *tsC(447)* mutation. This result demonstrated that all the *ts*⁺ clones tested from the persistently infected cells had undergone intragenic reversion events relative to the parental *tsC(447)* mutation. However, the fact that clones 9 and 26 yielded *ts* progeny on backcross to wild type indicated that these persistent virus clones harbored suppressed *ts* mutations in groups other than group C. Reassortment analysis of the *ts* mutations rescued from clone 9 showed that they were representatives of reassortment group G. Further reassortment analysis showed that *ts* mutants representing groups B, G, and H were rescued by backcross from clone 26. This result indicated that, while clone 26 did not contain an extragenically suppressed *tsC(447)* mutation, it did contain suppressed mutations in three other reassortment groups (AHMED et al. 1980). These results were highly unusual and contrasted with previous results in which nonparental *ts* mutations had only been recovered from backcrosses that had also yielded the parental *ts* mutation (RAMIG and FIELDS 1979). The extraordinary finding that clone 26 contained three distinct *ts* lesions in suppressed form, so that the clone expressed *ts*⁺ phenotype, illustrated the need to distinguish clearly between genotype and phenotype when analyzing viruses generated under circumstances where mutation might be favored (AHMED et al. 1980).

4.2.3 Rapid Screening for Extragenic Suppression

When stocks of *ts* mutants are plated in a plaque assay at nonpermissive temperature (39 °C), two populations of plaques are often observed (CROSS and FIELDS 1976b). One population of plaques is large and clear and characteristically has sharp, well-defined edges. These "lytic" plaques are typical of wild type and represent revertants present in the stock. The other, majority population is characteristically small and faint with diffuse edges and represents limited plaque formation by *ts* mutants that are incompletely lethal at 39 °C. A rapid method of screening for extragenically suppressed pseudorevertants was devised which depends on the distinction between the lytic plaques made by revertants or wild type and the "leak" plaques made by incompletely lethal *ts* mutants (RAMIG and FIELDS

1977). The progeny from the cross of a suspected extragenically suppressed revertant (lytic plaques) and wild type (lytic plaques) is plated at 39 °C and, following plaque formation and staining with neutral red, the plaques are examined for the presence of "leak" plaques in the yield. The observation of "leak" plaques among the progeny is a strong indication that one or more *ts* mutations were rescued from the revertant parent in the backcross. This method does not yield information on the reassortment group of the *ts* mutation rescued, only indicating that rescue of *ts* mutations occurred. As controls in this assay, the suspected extragenically suppressed revertant, wild type, and the parental *ts* mutant should be self-crossed and the yields examined to confirm that the potential extragenically suppressed revertant and wild type did not yield "leak" plaques and that the parental *ts* mutant yielded "leak" plaques as the majority type. This screening procedure can be completed in 6 days as compared to 48 days required for the standard method of identifying *ts* mutations among the progeny of a backcross. In one study in which this screening method was applied and the results were confirmed by standard methods, 15 of 19 revertants derived from "leaky" parental *ts* mutants were correctly classified as extragenically suppressed pseudorevertants (RAMIG and FIELDS 1979), indicating that the method is useful as a rapid screening technique.

4.2.4 Suppression of Other Phenotypes

All of the studies of suppression in reovirus cited above involved suppression of *ts* mutations. Suppression of mutations with non-*ts* phenotypes has not been reported or carefully studied in the reovirus system. Although suppression of other mutant phenotypes has not been documented in reovirus, studies with other animal virus systems indicate that a diverse array of phenotypes can be genetically suppressed. Thus suppression should be considered a possibility regardless of the type of mutant phenotype under study.

5 Suppression of Mutant Phenotype in Other Animal Viruses

As noted in the introduction, genetic suppression of mutant phenotype has now been documented in a wide array of RNA and DNA genome animal viruses. The intent here is not to reiterate information already presented about suppression in reovirus but rather to illustrate situations in which suppression was encountered that were not discussed above. Some of these situations seem likely to occur in reovirus as additional studies are performed.

5.1 Suppression Mediated By Protein–Nucleic Acid Interactions

Extragenic suppression of mutations in reovirus could be explained by a compensating protein interactions model. The following example, drawn from the

DNA genome SV40 virus, illustrates that similar suppressive activity of mutations can be manifested through interactions between protein and nucleic acid (SHORTLE et al. 1979). Site-directed single base pair mutations were introduced at the origin of replication (*ori*). The *ori* mutations abolished a *Bgl*I site within *ori* and rendered the DNA replication defective. The DNA containing the *ori* mutations were subjected to random mutagenesis and were then transfected into monkey cells. Plaques resembling those of wild type were selected and tested for the persistence of *Bgl*I resistance, indicating that they still contained the *ori* mutation or mutations. These plaques represented pseudorevertants of the original *ori* mutations. The pseudorevertants synthesized DNA at wild-type levels, indicating that the suppressor mutation was masking the replication-defective phenotype of the mutation in *ori*. The second site (suppressor) mutations were mapped by marker rescue techniques to the gene encoding T-antigen. T-antigen was previously known to interact with *ori* sequences in vitro, and this work demonstrated that the interaction had functional significance in vivo.

5.2 Suppression of Deletion Mutations

Deletion mutations have generally been considered more stable than mutations of other types because they generally did not revert or reverted at extremely low frequency. A recent study in influenza virus demonstrated that not only can deletion mutations revert, but they can revert by intragenic suppression mechanisms (TREANOR et al. 1991). The *ts143-1* mutant of influenza A/Alaska/77 is temperature sensitive and forms small plaques by virtue of a 36-nucleotide deletion in the NS genome segment which leads to an in-frame deletion of 12 amino acids in the NS1 protein. ts^+ revertants of *ts143-1* could be easily selected when the mutant was plated at nonpermissive temperature. To determine the mechanism of reversion, a revertant of *ts143-1* was backcrossed to wild-type virus; *ts* progeny plaques were isolated from among the progeny of the backcross, but grouping crosses showed that the *ts143-1* mutation was not rescued in the backcross. The failure to rescue *ts143-1* indicated that the revertant had reverted by some intragenic mechanism. Sequences of the NS segment from a *ts143-1* revertant and the parental *ts143-1* mutant were determined. The parental *ts143-1* and the *ts143-1* revertant contained the 36-base deletion, and the revertant contained an additional U to C substitution at position 94, which resulted in a Val to Ala substitution at position 23 of the NS1 protein. The *ts143-1* mutation (deletion) appeared to have reverted by a point mutation elsewhere in the gene and thus to be an intragenically suppressed revertant. Protein-folding algorithms predict that the deletion mutation disrupts an α-helical region between amino acids 51 and 84 of NS1 and the suppressor mutation introduces a region of α-helix between amino acids 18 and 27. Whether these changes represent compensating structural changes in NS1 that restore function in NS1 is unknown (TREANOR et al. 1991). Because of their perceived stability, deletion mutations were being considered as attenuating mutations in genetically engineered vaccines. These results suggest, however, that deletion mutations may not be as stable as thought, and the use of deletions as attenuating mutations is being reevaluated.

Similar results have been reported in SV40 virus, in which large deletions (79–187 bp) in the agnogene can be suppressed by second site suppressor mutations in the VP1 protein (BARKAN et al. 1987).

5.3 Suppression of Biological Phenotypes

A number of cases have been documented in which suppression was shown to mask phenotypes of biological relevance in viruses (e.g., epitope expression, attenuation/virulence). Two of these cases will be discussed briefly here.

An attenuated influenza virus vaccine candidate strain (A/Alaska/77-*ts1A2*) was developed by the incorporation of *ts* mutations in two different genes (polymerase subunits P1 and P3) of the virus strain (TOLPIN et al. 1981). When A/Alaska/77-*ts1A2* was administered to volunteers in a field trial, one child shed *ts*+ revertant virus although the child was not ill. The revertant isolate was called FV1319. To determine the mechanism by which the vaccine strain had reverted, FV1319 was backcrossed to wild type and progeny clones were analyzed. Forty-four of 179 progeny plaques analyzed had *ts* phenotype, indicating that FV1319 was an extragenically suppressed revertant. Grouping crosses with the rescued *ts* mutations showed that the reversion was complex, because none of the rescued *ts* mutations were in P1 and all of the rescued *ts* mutations were in P3. This indicated that the *ts* mutation in P1 had reverted by intragenic events (true reversion, pseudoreversion, intragenic suppression), while the *ts* mutation in P3 was extragenically suppressed (TOLPIN et al. 1981). Subsequent tests of the FV1319 revertant in volunteers and in animals indicated that it had coreverted the *ts* and attenuated phenotypes (TOLPIN et al. 1982). In this case, the fact that attenuation was accomplished by *ts* mutations allowed the rapid definition of the mechanism of reversion because the *ts* phenotype served as an in vitro proxy for the relevant attenuated phenotype in vivo. These studies resulted in a reevaluation of the use of *ts* mutations as attenuating mutations in vaccines.

The second example of suppression of a biologically significant phenotype also involved reversion of an attenuated strain of virus to virulence, in this case in Sindbis virus (SCHOEPP and JOHNSTON 1993). A number of mutations that affect the mouse virulence of Sindbis virus have been identified in the 5′ halves of the E1 and E2 glycoprotein genes. In the E2 gene, sites of particular importance appeared to reside at amino acid residues 62, 96, 114, and 159. To investigate the basis of virulence, a panel of viruses was generated with different amino acid substitutions at positions 62, 114, and 159 of E2 by site-directed mutagenesis. Screening the panel of mutants revealed that introduction of a positively charged amino acid (Ser to Arg) at position 114 produced the most pronounced attenuating effect of any of the single mutations at this position. Either Lys to Glu substitution at position 159 or Asn to Asp substitution at position 62 in the context of Arg at position 114 resulted in suppression of attenuation. Whether the attenuating effects of the mutations at positions 62 and 159 resulted from restoration of virulent conformation of E2 was unclear, as it was noted that either the Glu-159 or Asp-62 mutations alone on a wild-type background significantly increased virulence. These results did, however, clearly demonstrate that Lys to Glu substitution at position 159 or Asn to Asp substitution at position 62 could intragenically suppress the effects of an attenuating Ser to Arg substitution at position 114.

6 Concluding Remarks

The discussion presented in this chapter indicates that the genetic phenomena of reversion and suppression occur in animal viruses in general, and reovirus in particular, just as they do in other genetic systems. All four of the possible pathways from mutant phenotype to wild-type phenotype (true reversion, pseudoreversion, intragenic suppression, and extragenic suppression) were documented in animal viruses, emphasizing the commonality of genetic function of the various biologic systems.

It is clear that, when working with mutations, reversion of those mutations is likely. Reversion by any of the pathways may occur with mutants displaying any phenotype. The studies described generally made use of the easily manipulated *ts* type of conditionally lethal mutation. The knowledge that reversion can occur by a number of pathways should prepare the careful investigator to deal with unusual or unexpected results, even when working with less easily manipulated and demonstrated phenotypes.

Acknowledgements. This chapter is dedicated to the memory of Bernie Fields, in whose laboratory the author initiated his studies of suppression of viral phenotype.

References

Ahmed R, Graham AF (1977) Persistent infection in L cells with temperature-sensitive mutants of reovirus. J Virol 23:250–262

Ahmed R, Chakraborty PR, Graham AF, Ramig RF, Fields BN (1980) Genetic variation during persistent reovirus infection: presence of extragenically suppressed temperature-sensitive lesions in wild type virus isolated from persistently infected cells. J Virol 34:383–389

Barkan A, Welch RC, Mertz JE (1987) Missense mutations in the VP1 gene of simian virus 40 that compensate for defects caused by deletions in the viral agnogene. J Virol 61:3190–3198

Benzer S (1957) The elementary units of heredity. In: McElroy WD, Glass B (eds) The chemical basis of heredity. Hopkins, Baltimore, Maryland, pp 70–93

Campbell A (1961) Sensitive mutants of bacteriophage λ. Virology 14:22–32

Coombs KM, Mak S-C, Petrycky-Cox LD (1994) Studies of the major reovirus core protein σ2: reversion of the assembly-defective mutant tsC(447) is an intragenic process and involves back mutation of Asp-383 to Asn. J Virol 68:177–186

Cooper PD (1968) A genetic map of poliovirus temperature-sensitive mutants. Virology 35:584–594

Crick HFC, Barnett L, Brenner S, Towbin-Watts RJ (1961) General nature of the genetic code for proteins. Nature (Lond) 192:1227–1232

Cross RK, Fields BN (1972) Temperature-sensitive mutants of reovirus type 3: Studies on the synthesis of viral RNA. Virology 50:799–809

Cross RK, Fields BN (1976a) Temperature-sensitive mutants of reovirus type 3: evidence for aberrant μl and μ2 polypeptide species. J Virol 19:174–179

Cross RK, Fields BN (1976b) Use of an aberrant polypeptide as a marker in three-factor crosses: further evidence for independent reassortment as the mechanism of recombination between temperature-sensitive mutants of reovirus type 3. Virology 74:345–362

Dermody TS, Schiff LA, Mibert ML, Coombs KM, Fields BN (1991) The S2 gene nucleotide sequences of prototype strains of the three reovirus serotypes: characterization of reovirus core protein σ2. J Virol 65:5721–5731

Edgar RS (1966) Conditional lethals. In: Cairns J, Stent GS, Watson JD (eds) Phage and the origins of molecular biology. Cold Spring Harbor Laboratory of Quantitative Biology, Cold Spring Harbor, NY, pp 166

Fields BN (1971) Temperature-sensitive mutants of reovirus: features of genetic recombination. Virology 46:142–148

Fields BN, Joklik WK (1969) Isolation and preliminary genetic and biochemical characterization of temperature-sensitive mutants of reovirus. Virology 37:335–342

Fields BN, Raine CS, Baum SG (1971) Temperature-sensitive mutants of reovirus type 3: defects in viral maturation as studied by immunofluorescence and electron microscopy. Virology 43:569–578

Floor E (1970) Interaction of morphogenetic genes of bacteriophage T4. J Mol Biol 47:293–306

Ghendon Y, Markushin S, Lisovskaya K, Penn CR, Mahy BWJ (1982) Extragenic suppression of a ts phenotype during recombination between ts mutants of two fowl plague virus strains with a ts mutation in gene 1. J Gen Virol 62:239–248

Gorini L (1970) Informational suppression. Annu Rev Genet 4:107–134

Gorini L, Beckwith JR (1966) Suppression. Annu Rev Microbiol 20:401–422

Hall JD, Almy RE (1982) Evidence for control of herpes simplex virus mutagenesis by the viral DNA polymerase. Virology 116:535–543

Hartman JE, Roth JR (1973) Mechanisms of suppression. Adv Genet 17:1–105

Holland J, Spindler K, Horodyski F, Grabau E, Nichol S, Van de Pol S (1982) Rapid evolution of RNA genomes. Science 215:1577–1585

Jarvik J, Botstein D (1975) Conditional-lethal mutations that suppress genetic defects in morphogenesis by altering structural proteins. Proc Natl Acad Sci USA 72:2738–2742

Jayasuriya AKA (1991) Molecular characterization of the reovirus M2 gene. PhD Thesis, Harvard University, Cambridge, MA

King AMQ, Slade WR, Newman JWI, McCahon D (1980) Temperature-sensitive mutants of foot-and-mouth disease virus with altered structural polypeptides. II. Comparison of recombination and biochemical maps. J Virol 34:67–72

Kruijer W, Nicolas JC, vanSchaik FMA, Sussenbach JS (1983) Structure and function of DNA binding proteins from revertants of adenovirus type 5 mutants with a temperature-sensitive DNA replication. Virology 124:425–433

Lai MM (1992) RNA recombination in animal and plant viruses. Microbiol Revs 56:61–79

Lee PWK, Hayes EC, Joklik WK (1981) Protein sigma 1 is the reovirus cell attachment protein. Virology 108:156–163

McCrae MA, Joklik WK (1978) The nature of the polypeptide encoded by each of the ten double-stranded RNA segments of reovirus type 3. Virology 89:578–593

McFadden G, Essani K, Dales S (1980) A new endonuclease restriction site which is at a locus of a temperature-sensitive mutation in vaccinia virus is associated with true and pseudorevertants. Virology 101:277–280

McPhillips TH, Ramig RF (1984) Extragenic suppression of temperature-sensitive phenotype in reovirus: Mapping suppressor mutations. Virology 135:428–439

Mustoe TA, Ramig RF, Sharpe AH, Fields BN (1978a) A genetic map of reovirus. III. Assignment of the double-stranded RNA positive mutant groups A, B, and G to genome segments. Virology 85:545–556

Mustoe TA, Ramig RF, Sharpe AH, Fields BN (1978b) Genetics of reovirus: identification of the dsRNA segments encoding the polypeptides of the mu and sigma size classes. Virology 89:594–604

Ramig RF (1980) Suppression of temperature-sensitive phenotype in reovirus: an alternative pathway from ts to ts[+] phenotype. In: Fields BN, Jaenisch R, Fox CF (eds) Animal virus genetics. Academic, New York, pp 633–642

Ramig RF, Fields BN (1977) Method for rapidly screening revertants of reovirus temperature-sensitive mutants for extragenic suppression. Virology 81:170–173

Ramig RF, Fields BN (1979) Revertants of temperature-sensitive mutants of reovirus: evidence for frequent extragenic suppression. Virology 92:155–167

Ramig RF, Fields BN (1983) Genetics of reoviruses. In: Joklik WK (ed) The reoviridae. Plenum, New York, pp 197–228

Ramig RF, Ward RL (1991) Genomic segment reassortment in rotavirus and other reoviridae. Adv Virus Res 39:163–207

Ramig RF, White RM, Fields BN (1977) Suppression of the temperature-sensitive phenotype of a mutant of reovirus type 3. Science 195:406–407

Ramig RF, Mustoe TA, Sharpe AH, Fields BN (1978) A genetic map of reovirus. II. Assignment of the double-stranded RNA negative mutant groups C, D, and E to genome segments. Virology 85:531–544

Ramig RF, Ahmed R, Fields BN (1983) A genetic map of reovirus: assignment of the newly defined mutant groups H, I, and J to genome segments. Virology 125:299–313

Schiff LA, Nibert ML, Co MS, Brown EG, Fields BN (1988) Distinct binding sites for zinc and double-stranded RNA in the reovirus outer capsid protein sigma 3. Mol Cell Biol 8:273–283

Schoepp RJ, Johnston RE (1993) Sindbis virus pathogenesis: phenotypic reversion of an attenuated strain to virulence by second-site intragenic suppressor mutations. J Gen Virol 74:1691–1695

Scholtissek C, Spring SB (1982) Extragenic suppression of temperature sensitive mutations in RNA segment 8 by replacement of different segments with those of other influenza A virus prototype strains. Virology 118:28–34

Sharpe AH, Ramig RF, Mustoe TA, Fields BN (1978) A genetic map of reovirus. I. Correlation of genome RNAs between serotypes 1, 2, and 3. Virology 84:63–74

Shortle DR, Margolskee RF, Nathans D (1979) Mutational analysis of the simian virus 40 replicon: pseudorevertants of mutants with a defective replication origin. Proc Natl Acad Sci USA 76:6128–6131

Sturdevant AH (1920) Vermilion gene and gynandromorphism. Proc Soc Exp Biol Med 17:70–71

Tolpin MD, Massicot JG, Mullinix MG, Kim HW, Parrott RH, Chanock RM, Murphy BR (1981) Genetic factors associated with loss of temperature-sensitive phenotype of the influenza A/Alaska/77-ts-1A2 recombinant during growth in vivo. Virology 112:505–517

Tolpin MS, Clements ML, Levine MM, Black RE, Saah AJ, Anthony WC, Cisneros L, Chanock RM, Murphy BR (1982) Evaluation of a phenotypic revertant of the A/Alaska/77-ts-1A2 reassortant virus in hamsters and in seronegative adult volunteers: further evidence that the temperature-sensitive phenotype is responsible for attenuation of ts-1A2 reassortant viruses. Infect Immun 36:645–650

Treanor JJ, Buja R, Murphy BR (1991) Intragenic suppression of a deletion mutant for the nonstructural gene of an influenza virus. J Virol 65:4204–4210

Wiener JR, Joklik WK (1988) Evolution of reovirus genes: a comparison of serotype 1, 2, and 3 M2 genome segments which encode the major structural capsid protein μ1C. Virology 163:603–613

Wiener JR, McLaughlin T, Joklik WK (1989) The sequences of the S2 genome segments for reovirus serotype 3 and of the dsRNA-negative mutant ts447. Virology 170:340–341

Reovirus Cell Attachment Protein σ1: Structure–Function Relationships and Biogenesis

P.W.K. Lee and R. Gilmore

1 Introduction

The reovirus σ1 protein is encoded by the S1 gene segment and is probably the most extensively studied of all reovirus proteins. This is mainly because σ1 is the reovirus cell attachment protein and, as such, represents the first viral protein the cell encounters during reovirus invasion. As fibrous structures extending from the 12 vertices of the viral icosahedron, σ1 allows ready access of the virus to host cell receptors as well as multivalent binding, which is important for the subsequent virus internalization step. While the cell-binding function of σ1 can be readily demonstrated in vitro, the in vivo functions of σ1 have been deduced mainly from studies of animals infected with genetic reassortants derived from the three reovirus serotypes. It is interesting to note that most of the in vivo functions ascribed to σ1 can be explained by its cell-binding activity.

Department of Microbiology and Infectious Diseases, University of Calgary, Calgary, Alberta, Canada T2N 4N1

Important information on the structural organization of σ1 has been obtained from sequence analysis and electron microscopy, while the structure–function relationships of σ1 have been probed by molecular and by biochemical and biophysical means. Such studies have been facilitated by the fact that σ1 proteins expressed in both prokaryotic and eukaryotic systems, as well as in vitro, are all functional (i.e., capable of binding to cells). The expressed σ1 proteins from bacteria, baculovirus, and vaccinia virus recombinants, as well as purified σ1 from reovirions, have been characterized structurally and functionally, providing important information on σ1 architecture. The use of in vitro-synthesized σ1, in particular, has led to interesting revelations pertaining to its biogenesis; it is now an important model for the study of assembly and folding of cytosolic oligomers.

2 Functions

2.1 Cell Binding

The first indication that protein σ1 may be responsible for reovirus attachment came from genetic reassortment studies which showed that the reovirus S1 gene (encoding σ1) encodes the viral hemagglutinin (WEINER et al. 1978). Direct evidence that σ1 is the reovirus cell attachment protein was subsequently obtained and was based on the observation that, of all the soluble reovirus-specified proteins present in the infected cell lysate, protein σ1 alone possesses the capacity to bind to host cells (LEE et al. 1981). This binding is specific since it is blocked in the presence of excess reovirus. Binding competition studies have also revealed that the σ1 proteins of all three reovirus serotypes (type 1 Lang, type 2 Jones, and type 3 Dearing) compete for the same binding sites on mouse L fibroblasts.

Accumulated evidence suggests that the major signal on the cell surface initially recognized by σ1 is sialic acid. First, prior treatment of host cells with neuraminidase abrogates both reovirus binding and infection (GENTSCH and PACITTI 1985, 1987). Second, reovirus binding to host cells is inhibited in the presence of a high concentration of free N-acetylneuraminic acid and N-acetylneuraminyllactose, but not by free lactose (GENTSCH and PACCITI 1985; PAUL et al. 1989). Third, α-sialic acid alone (linked to bovine serum albumin) interacts directly with reovirus in a solid-phase binding system (PAUL et al. 1989). Since many glycoproteins on the cell surface are sialylated, it was theorized that reovirus should interact with a variety of cell surface structures, rather than with a single homogeneous species. This was indeed found to be the case from ligand blot analyses in which cell membrane proteins resolved by sodium dodecyl sulfate-polyacrylamide gel electrophoresis (SDS-PAGE) and blotted onto nitrocellulose were probed with labeled reovirus (CHOI et al. 1990; CHOI 1994). There is now little doubt that the initial interaction between reovirus and the host cell involves the binding of the σ1 globular head to sialic acid on sialic acid-bearing structures on the cell surface.

Recently, it was observed that the introduction of the epidermal growth factor receptor (EGFR) into reovirus-resistant cells renders them susceptible to reovirus infection (STRONG et al. 1993). The concurrent demonstration that EGFR directly interacts with reovirus (TANG et al. 1993) has led to the speculation that the binding of reovirus to EGFR might be important for a productive infection. Through the use of other oncogene products downstream of EGFR, it is now known that the binding of reovirus to EGFR (most likely to sialic acid moieties on EGFR) represents a fortuitous event that is unrelated to the ensuing infection (STRONG and LEE 1996; J.E. STRONG et al., unpublished data).

The reovirus receptor on erythrocytes has been shown to be the M and N blood group antigen glycophorin A (PAUL and LEE 1987). There is evidence that the nature of interaction between σ1 and glycophorin is distinct from the sialic acid–σ1 interaction in the case of host cells and that, in addition to the globular head, a region in the fibrous portion of σ1 is also involved in this interaction (see below).

2.2 Other Functions Ascribed to Protein σ1

The receptor-binding function of σ1 is probably accountable, at least in part, for the observation that the S1 gene defines reovirus tissue tropism (WEINER et al. 1977, 1980a; WOLF et al. 1981, 1983; KAUFFMAN et al. 1982; SHARPE and FIELDS 1985; TYLER et al. 1985) and dictates the pathway of virus spread in the host (TYLER et al. 1986). Other functions attributed to σ1 have included the triggering of both humoral and cellular immune responses in the host (including the generation of cytotoxic T lymphocytes, suppressor T cells, and the development of delayed-type hypersensitivity; WEINER and FIELDS 1977; FINBERG et al. 1979, 1982; FONTANA and WEINER 1980; WEINER et al. 1980b), interaction with the host cell microtubules (SHARPE et al. 1982), inhibition of host cell DNA synthesis (SHARPE and FIELDS 1981; GAULTON and GREENE 1989; TYLER et al. 1996), and induction of apoptosis (TYLER et al. 1995, 1996). How these other functions of σ1 are related to its structure is unclear at present; discussions on structure–function relationships of σ1 will therefore be limited to those that pertain to the receptor-binding function of this protein.

3 Structure–Function Relationships

3.1 Overall Structural Organization

Amino acid sequence comparison shows that there is only 10%–40% similarity between the σ1 proteins of the three serotypes, making it the least conserved of all the reovirus proteins (CASHDOLLAR et al. 1985; DUNCAN et al. 1990; NIBERT et al. 1990). This relative lack of similarity is accountable for the observation that σ1 is the principle antigen against which type-specific neutralizing antibodies are directed

(WEINER and FIELDS 1977). However, not only function, but also shape and configuration, have been retained by these three proteins (LEE et al. 1981; BANERJEA et al. 1988; FURLONG et al. 1988; FRASER et al. 1990). Sequence analysis of the S1 genes (encoding 470, 462, and 455 amino acids in the case of type 1, type 2, and type 3 reovirus, respectively) has revealed that the N-terminal third of σ1 is highly α-helical (NAGATA et al. 1984; BASSEL-DUBY et al. 1985; DERMODY et al. 1990a; DUNCAN et al. 1990; NIBERT et al. 1990). The additional presence, in the same region, of an extended heptad repeat $(a\text{-}b\text{-}c\text{-}d\text{-}e\text{-}f\text{-}g)_n$, where a and d are characteristically apolar residues, further indicates the propensity of this region to adopt a coiled-coil, rope-like structure (BASSEL-DUBY et al. 1985). The middle third of σ1 is composed largely of β-sheets. The C-terminal third of σ1 does not possess any distinct pattern and is therefore predicted to assume a complex globular structure. Based on this information, it was first postulated by BASSEL-DUBY et al. (1985) that σ1 is an oligomer and that the N- and C-terminal portions represent the proximal and the distal ends, respectively, of protein σ1, with the N-terminal end anchored to the virion and the C-terminal end interacting with the cell receptor. This theoretically deduced morphology of σ1 was subsequently confirmed by electron microscopy studies which showed purified σ1 as a rod-like structure (fiber) topped with a knob (BANERJEA et al. 1988; FURLONG et al. 1988; FRASER et al. 1990). It is more than 40 nm in length, with a tail 2–4 nm in diameter and a globular head approximately 8 nm in diameter. Similar lollipop-shaped structures have also been found on the surface of the virion, with the globular head region most distal from the virion (FURLONG et al. 1988). Thus the reovirus σ1 protein morphologically resembles the adenovirus fiber protein, which is also responsible for the attachment of the virus to cellular receptors. There is evidence, however, that σ1 can exist in two structural states in the virion: the extended form and a poorly characterized, more compact form. It has been suggested that the transition between the two states might be part of the infection process (FURLONG et al. 1988). However, only the extended state is considered here, since this is the functional state that purified σ1 assumes.

3.2 Trimeric Nature

Based on the icosahedral distribution of σ1 on the virion and on earlier stoichiometric data, it was initially suggested that two molecules of σ1 were present on each vertex of the virion (LEE et al. 1981). Early sequence analysis (BASSEL-DUBY et al. 1985) and characterization of purified σ1 (YEUNG et al. 1987) also appeared to be compatible with a dimeric model of σ1. Subsequent studies, including SDS-PAGE analysis of nondissociated σ1, chemical cross-linking of purified σ1, sequence analysis, electron microscopy image averaging, and dimensional analysis of purified fibers, apparently have supported a tetrameric model for the σ1 protein (BASSEL-DUBY et al. 1987; BANERJEA et al. 1988; BANERJEA and JOKLIK 1990; FRASER et al. 1990; NIBERT et al. 1990). On the other hand, analysis of the N-terminal half of σ1 by SDS-PAGE under nondissociating conditions suggests that it forms a three-stranded coiled coil (LEONE et al. 1991a).

The oligomeric status of σ1 has been resolved in our laboratory by two independent methods, one biochemical and the other biophysical (STRONG et al. 1991). The *biochemical* method involved the cosynthesis of the full-length protein (designated A) and a C-terminal truncated protein (designated B) in an in vitro translation system and the subsequent analysis of the products by SDS-PAGE under nondissociating conditions, which allowed oligomeric forms of σ1 to be identified. A total of four oligomeric bands were invariably observed. Two of the bands migrated at the same positions as the individually synthesized full-length and truncated proteins, whereas the middle two bands migrated with intermediate mobilities. The formation of four oligomeric protein species with different migration rates is consistent with the notion that σ1 is composed of three subunits, with the four bands corresponding to A_3, A_2B_1, A_1B_2, and B_3, respectively.

The *biophysical* method involved the analysis of purified σ1 by column filtration and sucrose gradient sedimentation. These two methods provided information on the Stokes radius and the sedimentation coefficient of σ1, respectively. The molecular mass of σ1 was then calculated to be 132 kDa according to the method presented by SIEGEL and MONTY (1966), making σ1 (monomeric molecular mass, 49 kDa) a trimer. We also took advantage of the fact that, under controlled conditions, trypsin cuts σ1 approximately in the middle, generating the N-terminal fibrous tail (monomeric molecular mass, 26 kDa) and the C-terminal globular head (monomeric molecular mass, 23 kDa). Biophysical analysis of the two fragments yielded molecular masses of 77 kDa and 64 kDa, respectively, both again corresponding to trimers. As expected, the N-terminal tryptic fragment is highly asymmetrical, with an f/f_o value of 1.92. The C-terminal tryptic fragment, on the other hand, is globular in nature, as indicated by its calculated frictional ratio of 1.37, a value that is comparable to that for the globular protein bovine serum albumen (f/f_o, 1.35).

3.3 Receptor-Binding Domains

The host cell receptor-binding domain of σ1 has been mapped to the C-terminal half of this protein. This has been achieved by two independent methods, one biochemical and the other molecular. The *biochemical* method involved the use of the two tryptic fragments (see above) for direct cell-binding assay, and only the 23-kDa fragment was found to bind (YEUNG et al. 1989). The 23-kDa fragment represents the C-terminal half of σ1, since it contains the single cysteine residue (amino acid 351) as revealed by tryptic analysis of [^{35}S]cysteine-labeled σ1. The binding of this 23-kDa fragment to cell receptors was found to be as efficient as that of the full-length σ1 protein.

The *molecular* approach of identifying the host cell receptor-binding domain of σ1 involved the analysis of a number of σ1 deletion or single-amino acid substitution mutants expressed in *Escherichia coli* (NAGATA et al. 1987; TURNER et al. 1992) or in vitro in rabbit reticulocyte lysates (DUNCAN et al. 1991). It was found that the deletion of as few as four amino acids from the C terminus totally abrogates σ1 cell-binding activity (DUNCAN et al. 1991). A similar loss of binding function was also observed when amino acids conserved among the σ1 proteins of the three reovirus serotypes and located at the C-terminal third of σ1 were sub- stituted (TURNER et al. 1992). Abrogation of cell-binding activity is invariably associated with a drastic conformational change in the C-terminal portion of σ1, based on protease sensitivity and monoclonal antibody recognition assays. Thus the σ1 receptor-binding site is composed of noncontiguous residues at the C ter- minus and, as such, represents a conformational, rather than a linear domain, as previously suggested (BRUCK et al. 1986; WILLIAMS et al. 1988). This is compatible with the observation that only a trimeric, properly assembled C-terminal head is capable of binding to cell receptors (LEONE et al. 1991c, 1992). It is not known, however, whether each subunit of the trimeric head harbors its own receptor (sialic acid)-binding domain or whether all three subunits together contribute to a single receptor-binding site. Limited deletion at the N terminus does not appear to affect the receptor-binding function of σ1. However, as discussed below, the N-terminal half of σ1 plays a crucial role in C-terminal trimerization, and hence the generation (but not the maintenance) of a fully functional C-terminal receptor-binding do- main, although it itself does not directly interact with the cellular receptor. It is interesting to note in this regard that mutations in this region of σ1, which affects σ1 stability, have been linked to the establishment of persistent reovirus infection in mouse L fibroblasts (WILSON et al. 1996).

The domain on σ1 involved in erythrocyte (glycophorin) binding is apparently not identical to that involved in host cell receptor (sialic acid) binding. Earlier studies using various monoclonal anti-σ1 antibodies revealed distinct epitopes on σ1 for these two functions (BURSTIN et al. 1982; SPRIGGS et al. 1983). The use of σ1 deletion mutants further mapped these domains to distinct regions on σ1 (NAGATA et al. 1987). In particular, the region immediately preceding the C-terminal globular head (residues 121–221) does not appear to play a major role in host cell receptor binding, but is crucial for glycophorin binding. This is also corroborated by the observation that the C-terminal tryptic fragment which binds to host cell receptors efficiently manifests no glycophorin-binding activity; neither does the N-terminal tryptic fragment (YEUNG et al. 1987). Furthermore, an examination of three type 3 field isolates revealed that strains negative for hemagglutination and glycophorin binding all possessed mutations within an N-terminal domain of the stem which was predicted to possess primarily β-sheet conformation in wild-type σ1 (DERMODY et al. 1990b). Collectively, these observations suggest that, whereas host cell re- ceptor binding involves strictly the C-terminal globular head, erythrocyte (glyco- phorin) binding involves both the globular head and part of the fibrous stem.

3.4 Virion-Anchoring Domain

The virion anchoring domain of σ1 is thought to be located at the extreme N terminus of the σ1 trimer. Several lines of evidence support this contention. Se- quence analysis reveals a short region of clustered hydrophobic amino acids from positions 1 to 12 in type 3 σ1 predicted to be α-helical (NIBERT et al. 1990; LEONE et al. 1991b). Their hydrophobicity suggests that they could be buried in the viral protein coat. This hydrophobic tail is followed by a short region of seven to ten residues predicted to form β-turns. Immediately downstream of this, the heptad repeat of the coiled-coil motif begins. The short β-turn sequence falls in a region which, judging from electron micrographs, appears to be highly flexible and has been postulated to act as a hinge between the hydrophobic tail and the coiled coil (FRASER et al. 1990; NIBERT et al. 1990), thereby allowing flexibility of the σ1 protein while anchored to the virus. This hinge region in type 3 σ1 is susceptible to chymotrypsin cleavage in isolated σ1, suggesting an open or loose conformation (LEONE et al. 1991b; DUNCAN and LEE 1994). This same susceptibility to chymotrypsin is not observed for virion-associated σ1, implying that this digestion site is hidden in the virus. The most convincing evidence that the extreme N terminus is the virion-anchoring domain comes from the demonstration that a chimeric protein containing the N-terminal quarter of type 3 σ1 fused to chloramphenicol acetyl-transferase (CAT) can be incorporated into the type 1 virus coat (MAH et al. 1990) as an oligomer (LEONE et al. 1991b). Deletion of the hydrophobic tail leads to the loss of incorporability into the virion. Interestingly, a short region of the heptad repeat (amino acids 38–107) is also an absolute requirement for incorporation (LEONE et al. 1991b). However, based on structural analysis and electron microscopy, it seems safe to conclude that the hydrophobic tail is directly involved in virion anchoring and that the short heptad repeat region is possibly important for the generation of an incorporation-competent tail. How the trimeric hydrophobic tail fits inside the lumen of the icosahedrally distributed pentameric λ2 spikes is unclear at present.

3.5 Demarcation of Structural and Functional Domains by Hinge Regions

It is interesting to note that electron microscopy of purified σ1 has revealed the presence of kinks in specific regions of the molecule, most notably near the N terminus, in the neck region immediately below the head, and near the middle of the fiber (FRASER et al. 1990). These sites correspond to potential hinge regions, as suggested by sequence-predicted flexibility profiles (NIBERT et al. 1990). Indeed, when purified type 3 σ1 is exposed to a variety of proteases with different cleavage specificities, cleavage invariably occurs in two specific regions in the molecule (DUNCAN and LEE 1994). These two regions are localized to the proposed N-terminal hinge region separating the hydrophobic anchor from the coiled coil and to the C-proximal portion of the neck separating most of the fibrous tail from the globular head. The positions of these sites relative to the known structural and functional domains of σ1 are depicted in Fig. 1. Although the significance of these protease-sensitive hinge regions is unclear at present, in view of the requirement of multivalent binding for viral entry, one can envisage how binding efficiency could be drastically enhanced for a virus with flexible arms compared to one with rigid extensions.

C-terminus

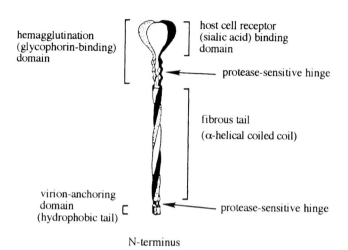

hemagglutination
(glycophorin-binding)
domain

host cell receptor
(sialic acid) binding
domain

protease-sensitive hinge

fibrous tail
(α-helical coiled coil)

virion-anchoring
domain
(hydrophobic tail)

protease-sensitive hinge

N-terminus

Fig. 1. General structural and functional domains of protein σ1

It deserves mention that, although the σ1 proteins of all three serotypes possess the aforementioned hinge regions, they are differentially susceptible to protease digestion. Specifically, type 1 σ1 is significantly more resistant to proteases than is type 3 σ1 (YEUNG et al. 1989; DUNCAN et al. 1994; NIBERT et al. 1995), and this in turn explains the differences in the capacity of type 1 and type 3 viruses to grow in intestinal tissue in the host (NIBERT et al. 1995).

4 Postattachment Dynamics

It is a well-established fact that viral attachment proteins undergo conformational changes upon binding to host cell receptors. These changes are necessary in order that subsequent events pertaining to viral entry can ensue. In the case of the reovirus σ1 protein, binding to host cell receptors (or sialic acid) leads to a drastic conformational change that is mapped, by deletion mutagenesis, to a region proximal to the N-terminal virion-anchoring domain (FERNANDES et al. 1994). This conformational change is completely reversible upon the release of σ1 from the cell surface. Importantly, when reovirions are used for binding, conformational changes in viral capsid proteins other than σ1 can also be detected. Specifically, the capsid proteins of cell-bound virions are more resistant to pepsin digestion. Since these changes are also observed using glutaraldehyde-fixed cells or plasma membranes instead of live cell, virus internalization is not responsible for this effect. As is the case for σ1, these conformational changes are reversible, since bound virions revert back to the pepsin-sensitive state upon release from the cell surface. Collectively, these results suggest that receptor (sialic acid) binding leads to a temporal progression of structural alteration in the virion, starting from the C-terminal receptor-binding domain of σ1, down the shaft of the protein toward the N-terminal virion-anchoring domain, and being subsequently relayed to the rest of the virion. These observations are also compatible with the data from cryoelectron microscopy and image reconstruction that reovirions and in vitro-generated intermediate subviral particles and cores (the latter two being accepted as true intermediates in reovirus infection) display significant differences in supramolecular structure and protein conformation that are related to the early steps of reovirus infection (DRYDEN et al. 1993). Of the two major outer-capsid proteins (μ1C and σ3), the μ1C protein, but not σ3, becomes particularly resistant to pepsin upon receptor binding (FERNANDES et al. 1994). Thus receptor-mediated conformational change in μ1C is probably important for reovirus entry. It is interesting to note in this regard that the μ1 protein has been associated with the capacity of reovirus to permeabilize host cell membranes (NIBERT and FIELDS 1992; HOOPER and FIELDS 1996a, b).

The observation that the receptor-induced conformational changes are reversible upon release of σ1 or of the virus from the receptor is interesting and is compatible with the concept that the binding that leads to reovirus entry is a two-step process involving distinct viral and cellular components, with the second step being a consequence of the first. Initial binding of σ1 to the ubiquitous sialic acid leads to conformational changes in the virion that are most likely a prerequisite for the second interaction. Under conditions that do not favor this latter step (e.g., due to inaccessibility of the altered virion to the pertinent cell component involved), the virus may detach from the sialic acid, revert back to its original conformation, and be ready to attach again. This process is repeated until conditions that favor the second interaction are met, and viral entry ensues as a consequence. It can easily be seen how such a "ping-pong" binding mechanism, which rejuvenates inconsequential encounters, could enhance the probability of productive infection.

5 Protein σ1 as a Model for Protein Assembly and Folding in the Cytosol

The study of the oligomeric status of σ1 has led to interesting revelations in general concepts pertaining to protein folding and oligomerization mechanisms. Indeed, protein σ1 lends itself perfectly for such studies for the following reasons. First, σ1 synthesized in an in vitro translation system (e.g., rabbit reticulocyte lysate) is fully functional (i.e., capable of binding to cells and recognizable by the mature σ1-specific monoclonal antibody G5; DUNCAN et al. 1991; LEONE et al. 1992; TURNER et al. 1992), and therefore parameters that could potentially perturb σ1 folding and trimerization (and hence function) are readily testable. Second, trimeric σ1 is stable in SDS-polyacrylamide gel under nondenaturing conditions and migrates as such,

thereby providing a simple and reliable means of following the trimerization process during σ1 biogenesis in vitro (LEONE et al. 1991a, c, 1992; TURNER et al. 1992).
Third, mature trimers (with assembled N and C termini) and immature trimers
(with assembled N terminus and *un*assembled C terminus) can be differentiated by
their relative migration rates in nondenaturing SDS-PAGE, allowing the characteristics and requirements unique to each oligomerization process (i.e., at the N or
C terminus) to be assessed (LEONE et al. 1992). These considerations have made σ1
an ideal model system for the study of folding and oligomerization of cytosolic
homooligomers.

5.1 Two Independent Trimerization Domains

Controlled trypsin digestion of σ1 results in the generation of two fragments corresponding to the N- and C-terminal halves of σ1, with both fragments retaining
their trimeric state and configuration (DUNCAN et al. 1991; STRONG et al. 1991). The
C-terminal tryptic fragment (amino acids 246–455) remains fully capable of binding
to cell receptors (YEUNG et al. 1989; DUNCAN et al. 1991). By incubating these
tryptic fragments in SDS-containing sample buffer at various temperatures followed by SDS-PAGE at 4 °C, it was found that the N-terminal trimer is considerably more stable than the C-terminal trimer, probably by virtue of its extended
α-helical heptad repeats (LEONE et al. 1991a). Moreover, when translated in vitro
independently as truncated proteins, the N-terminal half is capable of forming a
three-stranded α-helical coiled coil. On the other hand, the C-terminal half of σ1,
when expressed as a truncated protein (i.e., lacking the N-terminal half), does not
form trimers and manifests no cell-binding function (LEONE et al. 1992). Thus the
presence of the N-terminal half is important for the generation of a trimeric,
functional C-terminal globular head. Once this is accomplished, however, the N-
terminal half can be removed without affecting the structural or functional integrity
of the C-terminal half (DUNCAN et al. 1991). Collectively, these observations are
compatible with the view that trimerization of σ1 is initiated at the N terminus,
which is then followed by C-terminal trimerization with the accompanying conformational changes necessary for cell binding. That N-terminal trimerization
precedes C-terminal trimerization has indeed been demonstrated by a pulse-chase
experiment (Fig. 2; LEONE et al. 1996). Under SDS-PAGE conditions that allow for
the C-terminal head to remain trimeric, the gradual conversion of a slower
migrating σ1 species to one that migrates faster can be observed. The slower migrating species corresponds to immature σ1 trimers, which are hydra-like structures
with an assembled (trimeric) N-terminal half and an unassembled C-terminal half.
Subsequent trimerization of the C-terminal half results in a faster-migrating, lollipop-shaped structure. Of the two σ1 species, only the faster migrating, mature σ1
form possesses cell-binding function.

Although C-terminal trimerization is clearly dependent on N-terminal trimerization, there is evidence that the C-terminal half possesses its own assembly and
folding characteristics. Unlike N-terminal trimerization, which can tolerate a cer-

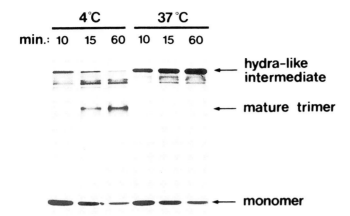

Fig. 2. Maturation of in vitro-synthesized σ1 trimers. Reovirus S1 transcripts were translated in rabbit reticulocyte lysate in the presence of [^{35}S]methionine for 10 min (pulse). Ribosomes were then pelleted and the supernatants, which contained no translation activity, were incubated at 37 °C for various durations (chase). Aliquots from both the pulse and chase samples were incubated in sodium dodecyl sulfate (SDS)-containing protein sample buffer at either 4 °C (at this temperature both N- and C-terminal trimers were stable) or 37 °C (at this temperature the N-terminal trimer, but not the C-terminal trimer was stable). SDS-polyacrylamide gel electrophoresis (PAGE) was carried out at 4 °C. (From LEONE et al. 1996)

tain degree of mutation, C-terminal trimerization is under stringent control. As mentioned above, deletion of as few as four amino acids from the C terminus totally abrogates the cell-binding function of σ1 (DUNCAN et al. 1991), as does the substitution of any of a number of conserved amino acids in the C-terminal region (TURNER et al. 1992). In both cases, the N-terminal half of σ1 remains intact, whereas the C-terminal half is grossly misfolded. Thus the mechanism of assembly and folding of the C-terminal half appears to be global in nature and is, in a way, independent of the N-terminal half of the protein. This hypothesis was subsequently proved by the demonstration that a deletion mutant lacking 100 residues in the N-terminal half immediately upstream of the C-terminal half of σ1 possesses a trimeric head and is fully capable of binding to cell receptors (LEONE et al. 1992; FERNANDES et al. 1994). Moreover, when such a mutant is cotranslated with full-length σ1, heterotrimers (corresponding to A_2B_1 and A_1B_2 described above) are formed which again possess properly folded heads (LEONE et al. 1992). In this case, part of the N-terminal half loops out to accommodate the formation of the trimeric head, demonstrating that perfect alignment of the three σ1 subunits throughout their entirety is not absolutely essential for this latter event. Thus, as far as C-terminal trimerization is concerned, N-terminal trimerization serves only to bring the three C-terminal subunits into close proximity to one another, thereby facilitating their subsequent trimerization.

5.2 Biogenesis

The demonstration that two independent trimerization domains exist within a single protein is significant because of its implications for the assembly scheme of cytosolic homooligomers. In the case of membrane-bound homooligomers such as the influenza virus hemagglutinin and the vesicular stomatitis virus (VSV) G protein, available evidence suggests that newly synthesized full-length proteins are anchored to the endoplasmic reticulum (ER) as partially folded monomers which subsequently undergo assembly with concomitant maturation folding (Doms et al. 1993). This post-translational assembly mechanism is unlikely to be adopted by cytosolic oligomers in view of the lack of any confining structure equivalent to the ER. For these proteins, such as σ1, cotranslational assembly with the initial interaction occurring between nascent chains on the polysome would represent a logical, if not the sole, alternative.

   To determine whether σ1 trimerization is initiated cotranslationally (i.e., at the ribosomal level), full-length σ1 and a C-terminally truncated σ1 protein were cosynthesized in vitro, and the products (homotrimers and heterotrimers) were analyzed by nondenaturing SDS-PAGE (Gilmore et al. 1996). The results indicate that there is invariably a preference for homotrimer over heterotrimer formation and that this bias is drastically enhanced when the levels of the two transcripts used in the reactions are proportionally reduced. This suggests that assembly at the N terminus occurs cotranslationally, a conclusion supported by the demonstration that trimeric σ1 chains are indeed present on S1 polysomes. The use of additional σ1 deletion mutants has further led to the deduction that this trimerization takes place when the ribosomes have traversed past the midpoint of the mRNA and is intrinsically an ATP-independent process. This latter observation is consistent with the lack of any demonstrable Hsp70-binding sites in the N-terminal third of σ1. It is not likely that N-terminal trimerization involves any chaperones.

   In contrast to N-terminal trimerization, the global nature of C-terminal trimerization predicts that it can proceed only when the C termini of all three subunits are intact and is accordingly a post-translational event (Leone et al. 1992). Indeed, hydra-like intermediates (with assembled N termini and unassembled C termini) in postribosomal fractions can be chased into mature σ1 forms with both the N and C termini in the trimeric state (Leone et al. 1996). Thus C-terminal trimerization differs from N-terminal trimerization in both temporality and stringency. Another feature that demarcates these two events is the clear involvement of ATP and Hsp70 (and possibly other chaperones) in the C-terminal trimerization process. Apyrase, which hydrolyses ATP and ADP, prevents globular head formation while having no effect on N-terminal triple coiled-coil formation (Gilmore et al. 1996; Leone et al. 1996). The requirement for ATP further implies the involvement of chaperones, including Hsp70. Indeed, Hsp70-binding sites were found to be present immediately downstream of the α-helical coiled coil as well as in the C-terminal half of σ1. Both nascent chains and the hydra-like intermediate, but not the mature σ1, are associated with Hsp70, whose role is the prevention of protein aggregation and improper assembly or folding. Recent evidence (R. Gilmore et al.,

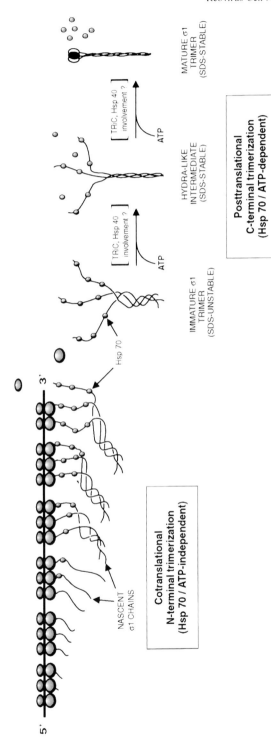

Fig. 3. Biogenesis model for the reovirus σ1 trimer. Assembly of the N-terminal triple-coiled coil occurs cotranslationally after the ribosomes have traversed past the midpoint of the polysome. This process is independent of ATP and Hsp70. Hsp70 binding to emerging residues downstream of the α-helical coiled-coil prevents aggregation and misfolding of the C terminus while sterically hindering tightening of the coiled coil. The immature trimer leaves the polysome as an sodium dodecyl sulfate (*SDS*)-sensitive structure still bound to Hsp70 and possibly other chaperones, such as Hsp40 and the TCP-1 ring complex (*TRiC*). Further ATP-dependent binding and release of chaperones lead first to the tightening of the coiled coil, generating an SDS-stable hydra-like intermediate, and then to global assembly and folding of the C terminus, generating a fully assembled lollipop-shaped mature σ1 trimer. (From LEONE et al. 1996)

unpublished data) suggests that, in addition to Hsp70, other chaperones such as Hsp90, Hsp40, and the chaperonin TCP-1 ring complex (TRiC) are also involved in C-terminal assembly of σ1.

The overall scheme of reovirus biogenesis is summarized in Fig. 3. Assembly of σ1 commences cotranslationally at the N terminus while the nascent chains are still attached to the actively translating ribosomes. This process does not involve Hsp70 or ATP and results in the generation of a loose triple coiled coil after the midpoint of the polysome, where Hsp70 (and possibly other chaperones) begins to interact with emerging residues. This prevents C-terminal misfolding and aggregation, but also hinders tightening of the coiled coil, and σ1 leaves the polysome as an SDS-unstable trimeric complex. Subsequent ATP-dependent release of the bound chaperones leads first to the tightening of the coiled coil and then to the second (C-terminal) trimerization event, which completes the σ1 maturation process.

6 Conclusion

The structure of the reovirus σ1 protein is uniquely suited to its function. The globular head serves to bind to the ubiquitous sialic acid moieties on the cell surface and the fibrous tail serves as an extension which facilitates this access. Rigidity and stability of the fibrous tail are provided by the extended triple coiled coil proximal to the short hydrophobic tail, which serves to anchor σ1 to the main body of the virus. The two hinge regions (both protease sensitive in the case of type 3 σ1) that separate the three major domains (globular head, fibrous tail, and anchoring domain) provide flexibility and promote multivalent binding, which is important for subsequent virus entry. There is cross-talk between the globular head and the fibrous tail since, upon binding to the cell receptor, a conformation change is relayed from the globular head via the fibrous tail to the main body of the virus. This is apparently necessary for a second binding step that eventually leads to virus internalization. With the structure–function relationships of σ1 having been defined in gross and general terms, what is urgently needed is the precise structure of σ1 (and of the virion) before and after receptor binding. This information will hopefully be provided by X-ray crystallography or cryoelectron microscopy studies of purified σ1 and/or the reovirion.

Research on σ1 biogenesis began as a "spin-off" of the structure–function studies, but has led to the interesting observation that σ1 possesses two independently active trimerization domains. Assembly of the N-terminal fibrous tail occurs cotranslationally and is independent of ATP/Hsp70, whereas formation of the C-terminal globular head is a post-translational event that requires both ATP and Hsp70. It remains to be seen whether other oligomeric proteins in the cytosol adopt a similar assembly strategy. Current research focuses on the involvement of other chaperones in σ1 biogenesis and the dynamics of this process. Knowledge gained from this exercise may in turn shed new light on the structure–function relationships of this protein.

References

Banerjea AC, Joklik WK (1990) Reovirus protein σ1 translated in vitro, as well as truncated derivatives of it that lack up to two-thirds of its C-terminal portion, exists as two major tetrameric molecular species that differ in electrophoretic mobility. Virology 179:460–462

Banerjea AC, Brechling KA, Ray CA, Erikson H, Pickup DJ, Joklik WK (1988) High-level synthesis of biologically active reovirus protein σ1 in a mammalian expression vector system. Virology 167:601–612

Bassel-Duby R, Jayasuriya A, Chatterjee D, Sonenberg N, Maizel JV Jr, Fields BN (1985) Sequence of reovirus hemagglutinin predicts a coiled-coil structure. Nature 315:421–423

Bassel-Duby R, Nibert ML, Homcy CJ, Fields BN, Sawutz DG (1987) Evidence that the sigma 1 protein of reovirus serotype 3 is a multimer. J Virol 61:1834–1841

Bruck C, Co MS, Slaoui M, Gaulton GN, Smith T, Fields BN, Mullins JI, Greene MI (1986) Nucleic acid sequence of an internal image-bearing monoclonal anti-idiotype and its comparison to the sequence of the external antigen. Proc Natl Acad Sci USA 83:6578–6582

Burstin SJ, Spriggs DR, Fields BN (1982) Evidence for functional domains on the reovirus type 3 hemagglutinin. Virology 117:146–155

Cashdollar LW, Chmelo RA, Wiener JR, Joklik WK (1985) Sequences of the S1 genes of the three serotypes of reovirus. Proc Natl Acad Sci USA 82:24–28

Choi AHC (1994) Internalization of virus binding proteins during entry of reovirus into K562 erythroleukemia cells. Virology 200:301–306

Choi AHC, Paul RW, Lee PWK (1990) Reovirus binds to multiple plasma membrane proteins of mouse L fibroblasts. Virology 178:316–320

Dermody TS, Nibert ML, Bassel-Duby R, Fields BN (1990a) Sequence diversity in S1 genes and S1 translation products of 11 serotype 3 reovirus strains. J Virol 64:4842–4850

Dermody TS, Nibert ML, Bassel-Duby R, Fields BN (1990b) A σ1 region important for hemagglutination by serotype 3 reovirus strains. J Virol 64:5173–5176

Doms RW, Lamb RA, Rose JK, Helenius A (1993) Folding and assembly of viral membrane proteins. Virology 193:545–562

Dryden KA, Wang G, Yeager M, Nibert ML, Coombs KM, Furlong DB, Fields BN, Baker TS (1993) Early steps in reovirus infection are associated with dramatic changes in supramolecular structure and protein conformation: analysis of virions and subviral particles by cryoelectron microscopy and image reconstruction. J Cell Biol 122:1023–1041

Duncan R, Lee PWK (1994) Localization of two protease-sensitive regions separating distinct domains in the reovirus cell-attachment protein σ1. Virology 203:149–152

Duncan R, Horne D, Cashdollar LW, Joklik WK, Lee PWK (1990) Identification of conserved domains in the cell attachment proteins of the three serotypes of reovirus. Virology 174:399–409

Duncan R, Horne D, Strong JE, Leone G, Pon RT, Yeung MC, Lee PWK (1991) Conformational and functional analysis of the C-terminal globular head of the reovirus cell attachment protein. Virology 182:810–819

Fernandes J, Tang D, Leone G, Lee PWK (1994) Binding of reovirus to receptor leads to conformational changes in viral capsid proteins that are reversible upon virus detachment. J Biol Chem 269:17043–17047

Finberg R, Weiner HL, Fields BN, Benacerraf B, Burakoff SJ (1979) Generation of cytolytic T lymphocytes after reovirus infection: role of the S1 gene. Proc Natl Acad Sci USA 76:442–446

Finberg R, Spriggs DR, Fields BN (1982) Host immune response to reovirus: CTL recognize the major neutralization domain of the viral hemagglutinin. J Immunol 129:2235–2238

Fontana A, Weiner HL (1980) Interaction of reovirus with cell surface receptors. II. Generation of suppressor T cells by the hemagglutinin of reovirus type 3. J Immunol 125:2660–2664

Fraser RDB, Furlong DB, Trus BL, Nibert BN, Fields BN, Steven AC (1990) Molecular structure of the cell-attachment protein of reovirus: correlation of computer-processed electron micrographs with sequence-based predictions. J Virol 64:2990–3000

Furlong DB, Nibert ML, Fields BN (1988) Sigma 1 protein of mammalian reoviruses extends from the surfaces of viral particles. J Virol 62:246–256

Gaulton GN, Greene MI (1989) Inhibition of cellular DNA synthesis by reovirus occurs through a receptor-linked signaling pathway that is mimicked by antiidiotypic, antireceptor antibody. J Exp Med 169:197–211

Gentsch JR, Pacitti AF (1985) Effect of neuraminidase treatment of cells and effect of soluble glyco-proteins on type 3 reovirus attachment to murine L-cells. J Virol 56:356–364

Gentsch JR, Pacitti AF (1987) Differential interaction of reovirus type 3 with sialylated receptor components on animal cells. Virology 161:245–248

Gilmore R, Coffey MC, Leone G, McLure K, Lee PWK (1996) Co-translational trimerization of the reovirus cell attachment protein. EMBO J 15:2651–2658

Hooper JW, Fields BN (1996a) Role of the μ1 protein in reovirus stability and capacity to cause chromium release from host cells. J Virol 70:459–467

Hooper JW, Fields BN (1996b) Monoclonal antibodies to reovirus σ1 and μ1 proteins inhibit chromium release from mouse L cells. J Virol 70:672–677

Kauffman RS, Wolf JL, Finberg R, Trier JS, Fields BN (1982) The σ1 protein determines the extent of spread of reovirus from the gastrointestinal tract of mice. Virology 124:403–410

Lee PWK, Hayes EC, Joklik WK (1981) Protein σ1 is the reovirus cell attachment protein. Virology 108:156–163

Leone G, Maybaum L, Lee PWK (1992) The reovirus cell attachment protein possesses two independently active trimerization domains: basis of dominant negative effects. Cell 71:479–488

Leone G, Duncan R, Mah DCW, Price A, Cashdollar LW, Lee PWK (1991a) The N-terminal heptad repeat region of reovirus cell attachment protein σ1 is responsible for σ1 oligomer stability and possesses intrinsic oligomerization function. Virology 182:336–345

Leone G, Mah DCW, Lee PWK (1991b) The incorporation of reovirus cell attachment protein σ1 into virions requires the N-terminal hydrophobic tail and the adjacent heptad repeat region. Virology 182:346–350

Leone G, Duncan R, Lee PWK (1991c) Trimerization of the reovirus cell attachment protein (σ1) induces conformational changes in σ1 necessary for its cell-binding function. Virology 184:758–761

Leone G, Coffey MC, Gilmore R, Duncan R, Maybaum L, Lee PWK (1996) C-terminal trimerization, but not N-terminal trimerization, of the reovirus cell attachment protein is a post-translational and Hsp70/ATP-dependent process. J Biol Chem 271:8466–8471

Mah DC, Leone G, Jankowski JM, Lee PWK (1990) The N-terminal quarter of reovirus cell attachment protein σ1 possesses intrinsic virion-anchoring function. Virology 179:95–103

Nagata L, Masri SA, Mah DCW, Lee PWK (1984) Molecular cloning and sequencing of the reovirus (serotype 3) S1 gene which encodes the viral cell attachment protein σ1. Nucleic Acids Res 12:8699–8710

Nagata L, Masri SA, Pon RT, Lee PWK (1987) Analysis of functional domains on reovirus cell attachment protein σ1 using cloned σ1 gene deletion mutants. Virology 160:162–168

Nibert ML, Fields BN (1992) A carboxy-terminal fragment of protein μ1/μ1C is present in infectious subvirion particles of mammalian reoviruses and is proposed to have a role in penetration. J Virol 66:6408–6418

Nibert ML, Dermody TS, Fields BN (1990) Structure of the reovirus cell-attachment protein: a model for the domain organization of σ1. J Virol 64:2976–2989

Nibert ML, Chappell JD, Dermody TS (1995) Infectious subvirion particles of reovirus type 3 Dearing exhibit a loss in infectivity and contain a cleaved sigma 1 protein. J Virol 69:5057–5067

Paul RW, Lee PWK (1987) Glycophorin is the reovirus receptor on human erythrocytes. Virology 159:94–101

Paul RW, Choi AHC, Lee PWK (1989) The α-anomeric form of sialic acid is the minimal receptor determinant recognized by reovirus. Virology 172:382–385

Sharpe AH, Fields BN (1981) Reovirus inhibition of cellular DNA synthesis: role of the S1 gene. J Virol 38:389–392

Sharpe AH, Fields BN (1985) Pathogenesis of viral infections. Basis concepts derived from the reovirus model. New Engl J Med 312:486–497

Sharpe AH, Chen LB, Fields BN (1982) The interaction of mammalian reoviruses with the cytoskeleton of monkey kidney CV-1 cells. Virology 120:399–411

Siegel LM, Monty KJ (1966) Determination of molecular weights and frictional ratios of proteins in impure systems by use of gel infiltration and density gradient centrifugation: application to crude preparations of sulfite and hydroxylamine reductases. Biochim Biophys Acta 112:346–362

Spriggs DR, Kaye K, Fields BN (1983) Topological analysis of the reovirus type 3 hemagglutinin. Virology 127:220–224

Strong JE, Lee PWK (1996) The v-erbB oncogene confers enhanced cellular susceptibility to reovirus infection. J Virol 70:612–616

Strong JE, Leone G, Duncan R, Sharma RK, Lee PWK (1991) Biochemical and biophysical characterization of the reovirus cell attachment protein σ1: evidence that it is a homotrimer. Virology
184:23–32

Strong JE, Tang D, Lee PWK (1993) Evidence that the epidermal growth factor receptor on host cells
confers reovirus infection efficiency. Virology 197:405–411

Tang D, Strong JE, Lee PWK (1993) Recognition of the epidermal growth factor receptor by reovirus.
Virology 197:412–414

Turner DL, Duncan R, Lee PWK (1992) Site-directed mutagenesis of the c-terminal portion of reovirus
protein σ1: evidence for a conformation-dependent receptor binding domain. Virology 186:219–227

Tyler KL, Bronson RT, Byers KB, Fields BN (1985) Molecular basis of viral neurotropism: experimental
reovirus infection. Neurology 35:88–92

Tyler KL, McPhee DA, Fields BN (1986) Distinct pathways of viral spread in the host determined by
reovirus σ1 gene segment. Science 233:770–774

Tyler KL, Squier MKT, Rodgers SE, Schneider BE, Oberhaus SM, Grdina TA, Cohen JJ, Dermody TS
(1995) Differences in the capacity of reovirus strains to induce apoptosis are determined by the viral
attachment protein s1. J Virol 69:6972–6979

Tyler KL, Squier MKT, Brown AL, Pike B, Willis D, Oberhaus SM, Dermody TS, Cohen JJ (1996)
Linkage between reovirus-induced apoptosis and inhibition of cellular DNA synthesis: role of the S1
and M2 genes. J Virol 70:7984–7991

Weiner HL, Fields BN (1977) Neutralization of reovirus: the gene responsible for the neutralization
antigen. J Exp Med 146:1305–1310

Weiner HL, Drayna D, Averill DR Jr, Fields BN (1977) Molecular basis of reovirus virulence: role of the
S1 gene. Proc Natl Acad Sci USA 74:5744–5748

Weiner HL, Ramig RF, Mustoe TA, Fields BN (1978) Identification of the gene coding for the hemagglutinin of reovirus. Virology 86:581–584

Weiner HL, Powers ML, Fields BN (1980a) Absolute linkage of virulence and central nervous system cell
tropism of reoviruses to viral hemagglutinin. J Infect Dis 141:609–616

Weiner HL, Greene MI, Fields BN (1980b) Delayed hypersensitivity in mice infected with reovirus.
I. Identification of host and viral gene products responsible for the immune response. J Immunol
125:278–282

Williams WV, Guy HR, Rubin DH, Robey F, Myers JM, Kieber-Emmons T, Weiner DB, Greene MI
(1988) Sequences of the cell-attachment sites of reovirus type 3 and its anti-idiotypic/antireceptor
antibody: modeling of their three-dimensional structures. Proc Natl Acad Sci USA 85:6488–6492

Wilson GJ, Wetzel JD, Puryear W, Bassel-Duby R, Dermody TS (1996) Persistent reovirus infections of
L cells select mutations in viral attachment protein σ1 that alter oligomer stability. J Virol 70:6598–
6606

Wolf JL, Rubin DH, Finberg R, Kauffman RS, Sharpe AH, Trier JS, Fields BN (1981) Intestinal M cells:
a pathway for entry of reovirus into the host. Science 212:471–472

Wolf JL, Kauffman RS, Finberg R, Dambrauskas R, Fields BN, Trier JS (1983) Determinants of reovirus
interaction with the intestinal M cells and absorptive cells of murine intestine. Gastroenterology
85:291–300

Yeung MC, Gill MJ, Alibhai SS, Shahrabadi MS, Lee PWK (1987) Purification and characterization of
the reovirus cell attachment protein σ1. Virology 156:377–385

Yeung MC, Lim D, Duncan R, Shahrabadi MS, Cashdollar LW, Lee PWK (1989) The cell attachment
proteins of type 1 and type 3 reovirus are differentially susceptible to trypsin and chymotrypsin.
Virology 170:62–70

Signal Transduction and Antiproliferative Function of the Mammalian Receptor for Type 3 Reovirus

H.U. Saragovi[1,2], N. Rebai[1], E. Roux[1], M. Gagnon[1], X. Zhang[3], B. Robaire[1], J. Bromberg[4], and M.I. Greene[3]

1 Background

Viruses exploit cellular proteins or membrane components as receptors or docking sites for binding and infection (reviewed in SARAGOVI et al. 1992a; SAUVÉ et al. 1992). A small list includes mammalian molecules such as epidermal growth factor receptors (EGFR), C3d receptors, CD4, EPO receptors, CD46, and interleukin

[1]Department of Pharmacology and Therapeutics, McGill University, 3655 Drummond Street, Montreal, QC, Canada H3G 1Y6

[2]McGill Cancer Centre, McGill University, 3655 Drummond Street, Montreal, QC, Canada H3G 1Y6

[3]University of Pennsylvania School of Medicine, Department of Pathology and Laboratory Medicine, 252 John Morgan Building, Philadelphia, PA 19104-6082, USA

[4]University of Michigan Medical Center, Department of Transplant Surgery, 2926 Taubman Center, Box 0331, Ann Arbor, MI 48109-0331, USA

(IL)-2 p75(β) receptors (BLOMQUIST et al. 1984; KLATZMANN et al. 1984; FRADE et al. 1985; LI et al. 1990). These mammalian cell surface molecules have well-known physiological functions, and their role as viral docking sites is pathological.

Teleologically, it seems advantageous for viruses to have selected critical proteins as docking sites. First, the virus is assured a stable entry site because mutations or downregulation of the docking proteins is likely to be deleterious. Second, the cognate function of the docking proteins can be subverted to induce events that may be advantageous for viral replication such as growth arrest, anti-apoptotic signals, or uncoupling from associated molecules. Therefore, using viruses as biological and molecular probes, novel molecules and functions have been identified.

2 Reovirus Infection

Reoviruses are double-stranded RNA viruses with a ubiquitous host range in mammals. Three serotypes (types 1–3) have been isolated from humans. The σ1 gene from reovirus encodes the hemagglutinins (HA), which define viral tropism and mediate many of the viral binding and biological processes (for a review, see TYLER and FIELDS 1990). HA1 (type 1), HA2 (type 2), and HA3 (type 3) bind to sialic acid-containing proteins in erythrocytes and cause hemagglutination, hence the name hemagglutinin (GENTSCH and PACITTI 1985; PACITTI and GENTSCH 1987; PAUL and LEE 1987).

A different domain of HA3 binds to the surface of neural and lymphoid cells (WEINER et al. 1980; LEE et al. 1981; TARDIEU et al. 1982; EPSTEIN et al. 1984). Cells of the central nervous system and the lymphoid system can be infected by type 3 reovirus (NEPOM et al. 1982), but not by type 1 reovirus expressing HA1 (KILHAM and MARGOLIS 1969; DICHTER et al. 1986). Reovirus type 3 is spread through peripheral or olfactory neurons to the central nervous system (GREENE et al. 1987; COHEN et al. 1990b). Intracerebral infection with reovirus type 3 results in acute encephalitis and 100% mortality (KILHAM and MARGOLIS 1969). Lymphoid cells have been shown to be infected in vitro by reovirus type 3, and it has been proposed that persistent infection plays a role in autoimmunity (MATSUZAKI et al. 1986; GAULTON and GREENE 1989; for a reviewed, see SIEGEL et al. 1990). These cells express specific "receptors" for HA3 which are of high affinity and saturable binding.

3 Initial Studies

More than a decade ago, Fields and colleagues observed that type 3 reoviruses inhibited host cell macromolecular synthesis. Inhibition of host cell DNA synthesis occurs 4–8 h after infection of epithelial cells, mitogen-stimulated T splenocytes, and other cell types. Fields and colleagues further observed that inhibition of cellular DNA synthesis was unrelated to viral replication because ultraviolet (UV)-killed virus caused the same antimitotic effect (reviewed in TYLER and FIELDS 1990). More recently, purified recombinant HA3 produced in bacteria was also used to bind receptors and to achieve antimitotic effects comparable to the virus (SARAGOVI et al. 1995).

Thus it was hypothesized that the biological effect of the reovirus type 3 or HA3 might be due to cell surface receptor perturbation rather than infection (TARDIEU et al. 1982; GAULTON and GREENE 1989; COHEN et al. 1990a). A "receptor" was functionally identified as a cell surface structure which bound intact reovirus type 3 or purified HA3. The cell surface structure was therefore named the reovirus type 3 receptor. However, it should be recognized that the viral receptor function is a pathological use of a cellular molecule and that the receptor has cognate functions of value to the cell.

4 Receptor-Binding Agents and Analogues

HA3 recognizes several sialic acid-containing surface proteins, making it difficult to determine conclusively which protein or proteins transduce the antimitotic signal. Anti-receptor monoclonal antibodies (mAb) were therefore produced as tools to study the reovirus type 3 receptor (ERTL et al. 1982; NEPOM et al. 1982; KAUFFMAN et al. 1983; NOSEWORTHY et al. 1983; SHARPE et al. 1984) using anti-idiotypic techniques.

4.1 Monoclonal Antibodies

The idiotype was the anti-HA3 mAb 9BG5, which neutralizes viral HA3. An anti-idiotypic mAb, designated 87.92.6, was found to be an internal image of HA3 (BRUCK et al. 1986). Binding of mAb 87.92.6 to a panel of murine and human cells was comparable to the binding of HA3; prior incubation with mAb 87.92.6 specifically inhibited the binding of type 3 reovirus (Fig. 1).

4.2 Receptor-Specific Small Molecules

The complementarity determining regions (CDR) of mAb 87.92.6 have sequence and structural homology with viral HA3; they endow the mAb with binding comparable to HA3. Small peptidic (WILLIAMS et al. 1988, 1989) and peptidomimetic (SARAGOVI et al. 1991, 1992b) analogues of the 87.92.6 CDR inhibit the interactions of HA3 or mAb 87.92.6 with mAb 9BG5 (Fig. 1). The small molecule

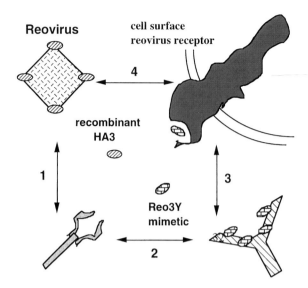

Fig. 1. Interactions between cell surface reovirus receptors and their ligands: *1*, monoclonal antibody (mAb) 9BG5 and reovirus or HA3; *2*, mAb 9BG5 and anti-idiotypic mAb 87.92.6; *3*, mAb 87.92.6 and a cellular receptor; *4*, reovirus or recombinant HA3 and a cellular receptor. The ligands mAb 87.92.6, mAb 87.92.6 peptidomimetics (termed Reo3Y), reovirus type 3, and recombinant HA3 all bind a cellular receptor (for simplicity shown as a single polypeptide) and the idiotypic mAb 9BG5. Reovirus type 3, HA3, mAb 87.92.6, and its peptidomimetic analogues inhibit cellular proliferation

analogues bind to a cell surface receptor which is also recognized by HA3 and mAb 87.92.6. Like their parent ligands and UV-killed reovirus, the small molecule analogues inhibit the DNA synthesis of cultured cells (SARAGOVI et al. 1991, 1995).

Therefore, the characterization of the signal-transducing mechanisms and bioactivity of the reovirus type 3 receptor offers an opportunity to identify novel elements or functions in the regulation of DNA synthesis. Importantly, the availability of small peptidic and peptidomimetic agents that bind to the receptor and inhibit proliferation may prove useful for the development of antimitotic or antitumor drugs (SARAGOVI et al. 1992b).

5 Biochemical Studies

We have studied the proteins specifically recognized by mAb 87.92.6. Biochemical isolation of the receptor complex by immunoprecipitation from normal tissues (brain, lymphoid) and from several cell lines with mAb 87.92.6 revealed two noncovalently associated polypeptides of 60–70 kDa (p65) (Co et al. 1985a, b; LIU et al. 1988) and 92–98 kDa (p95) (SARAGOVI et al. 1995; REBAI et al. 1996). This receptor complex has thus been termed p65/p95. HA3 and mAb 87.92.6 bind the p65 subunit, and the p95 subunit appears to be noncovalently associated and co-immunoprecipitated with p65.

The p65/p95 subunits isolated from T cell lines, neural cell lines, activated normal T lymphocytes, or normal fetal neural cells are biochemically indistinguishable by sodium dodecyl sulfate-polyacrylamide gel electrophoresis (SDS-PAGE), isoelectric focusing (IEF), and V8 peptide mapping (Rebai et al. 1996). However, the p65/p95 complex appears to be coupled to unique intracellular signaling pathways in different cells.

6 Phenotypic Studies of p65/p95 Receptor Expression

6.1 Mitotic cells

Phenotypic characterization of p65 receptor expression was performed by FAC-Scan analysis with mAb 87.92.6 in normal murine tissues and cell lines, followed by biochemical detection of p95 when p65 was expressed (p95 could not be studied in the absence of p65, since there are no reagents available at this time).

Resting peripheral T lymphocytes do not express p65 subunits, but mitogenic activation induces expression of high levels of surface p65/p95. Therefore, p65/p95 is an activation marker of T lymphocytes. Its induction closely follows early T cell division, and differentiation or activation markers (Saragovi et al. 1995).

In brain, p65/p95 expression precisely follows the active phase of neurogenesis. Receptor expression decreases developmentally as mouse neural cells lose proliferative potential and differentiate. A decrease in mAb 87.92.6 immunostaining is seen from embryonic day 15 (E15), to E21 (birth), to postnatal day 7 (PN7) (approximately 75%, approximately 25%, and less than 10% of all neural cells, respectively). In the cerebellum, approximately 60% of the cells express p65/p95 from E21 to PN7. Since the cerebellum is populated at this time, the data suggest that cells migrating in or proliferating in situ express the p65/p95 complex (Rebai et al. 1996).

6.2 Meiotic Cells

DNA synthesis and cell division also takes place in germ cells. In the testis, where there is rapid cellular proliferation, both mitotic and meiotic divisions occur. Each stem cell (spermatogonium) undergoes five mitotic divisions before undergoing two meiotic divisions (spermatocytes), producing approximately 120 million spermatozoa daily in man (Matsumoto 1996). Mitosis and meiosis share common features during the complex process of spermatogenesis. Thus immunocytochemical studies of sections of adult rat testes were undertaken with mAb 87.92.6 to assess whether p65/p95 expression is stage dependent in meiotic germ cells (Table 1).

Table 1. Monoclonal antibody (mAb) 87.92.6 immunocytochemistry of rat testis

Cell type	Stages					
	I–III	IV–VI	VII	VIII–X	XI, XII	XIII–XIV
Spermatogonia	+ + +	+ + + + +	–	+ +	+ + +	+ + +
Spermatocytes	+ + + + +	+ + + + +	+ + + +	+ + + +	+ + +	+ +
Spermatids	–	–	–	–	–	–

Staining was carried out in the presence of the idiotypic 9BG5 mAb, which binds and neutralizes mAb 87.92.6, demonstrating specificity. Further, control primary mAb did not reveal the same staining pattern. Over 1500 cells were staged and their label quantitated. Intensity of staining of the various cell types at each stage of spermatogenesis was identified as described by Dym and Clermont (1970).

In the testis, the two main populations of somatic cells (Leydig cells in the interstitium and Sertoli cells in the seminiferous tubules) were unlabeled, whereas germ cells showed a stage-specific labeling pattern (Roux et al., in press). In the interstitium, endothelial cells and macrophage-like cells were positive. In the seminiferous epithelium, specific germ cell populations were labeled; these included spermatogonia and all steps of primary and secondary spermatocytes, but not spermatids or spermatozoa. The labeled cells displayed different staining intensities in a stage-dependent manner. The label was mostly nuclear. During meiosis, the labeling in the nucleus was clearly associated with the chromosomes at anaphase, telophase, and cytokinesis (late telophase).

Thus, in seminiferous epithelium and interstitium of the adult rat testis, p65/p95 receptor expression is detected in dividing cells, both mitotic and meiotic. Nondividing cells such as Sertoli and Leydig cells, spermatids, and spermatozoa do not express p65/p95. The presence of p65/p95 receptors in the nucleus of meiotic cells in association with chromosomes suggests that this receptor may be involved in the regulation of meiosis. Further work will be necessary to understand the role of p65/p95 in the reproductive system.

Taken together, the data indicate that all immunostained cells within the testis, brain, or lymphoid tissues undergo nuclear decondensation, DNA synthesis, and cell division. It is an intriguing possibility that p65/p95 may regulate or be associated with these processes.

7 p65/p95-Mediated Signal Transduction

Binding of p65/p95 in intact cells activates a tyrosine kinase (TK) cascade rapidly downregulated by one or more tyrosine phosphatases (Saragovi et al. 1995; Rebai et al. 1996). The TK activity seems to be intrinsic to the p95 subunit. Thus, although HA3, mAb 87.92.6, and their peptidomimetic analogues bind the p65 component, the p95 subunit is activated. This feature is typical of many dimeric and multimeric complexes, where one subunit contributes an affinity-binding component and another a signaling component (reviewed in ULLRICH and SCHLESSINGER 1990; ULLMAN et al. 1990).

Activation of the p65/p95 TK by binding the receptor in intact cells causes de novo phosphotyrosinylation of specific intracellular substrates. In T cells, proteins with a molecular weight of 42 (pp42), 60 (pp60), 105 (pp105), and 185 (pp185) (SARAGOVI et al. 1995) are phosphotyrosinylated, whereas in neural cells proteins with a molecular weight of 60 (pp60), 80 (pp80), and 90 (pp90) are phosphotyrosinylated (REBAI et al. 1996). Phosphotyrosinylation is transient, and within 40 min the proteins return to baseline levels, possibly due to the activity of tyrosine phosphatases. Different substrate coupling in distinct tissues may lead to different signal transduction in T lymphoid versus neuronal cells.

8 p65/p95 Ligand-Mediated Antiproliferative Signal

We postulated that $p21^{ras}$ inactivation binding may be linked to the growth arrest mediated by p65/p95 binding. Thus we tested the hypothesis that constitutively active oncogenic v-Ha-ras would circumvent the antiproliferative effect.

The proliferation of R1.1 (thymoma) and P19 (embryonic teratocarcinoma) cells was monitored in response to p65/p95 binding. Susceptibility to growth inhibition was tested using either wild-type cells, cells after transfection with pZEM-neo vector control, or pZEM-neo containing the v-Ha-ras gene (Table 2). The transfected and wild-type cells express equivalent surface levels of p65/p95.

The growth of control wild-type or control pZEM-neo transfected cells were inhibited by mAb 87.92.6, by Reo3Y (a synthetic peptide analogue of mAb 87.92.6), and by recombinant HA3. Inhibition of proliferation ranged from

Table 2. Prevention of p65/p95-mediated growth inhibition by oncogenic $p21^{ras}$

Treatment	Proliferation (% of media)			
	R1.1-wt	R1.1-Ras	P19-wt	P19-Ras
Control mAb	106 ± 3	108 ± 4	110 ± 4	101 ± 5
87.92.6 mAb	38 ± 3	101 ± 2	50 ± 4	100 ± 1
HA3	69 ± 2	98 ± 3	63 ± 3	95 ± 1

pZEM-neo vectors with v-Ha-ras (SZYF et al. 1995) were stably transfected into R1.1 thymomas and P19 embryonic teratocarcinoma cells. Expression of transfected v-Ha-ras mRNA in various G418-resistant subclones was confirmed by northern blot analysis. FACScan analysis indicated that all clones expressed similar levels of p65/p95 receptors (data not shown). The indicated cells, i.e., wild-type (wt) or v-Ha-ras-expressing (Ras) cells, were grown in 96-well plates (approximately 7500 per well) in culture media containing 5% serum. Cells were either not treated or were treated with control anti-human p75(β) interleukin (IL)-2 receptor monoclonal antibody (mAb) 20G6 (10 nM), mAb 87.92.6 (10 nM), or purified recombinant HA3 (1 nM). After approximately 18 h, the status of the cells was quantitated using the MTT method. Data was standardized to untreated samples (100%) ± SEM. Each experiment was repeated at least three times (n = 4 for each cell line or transfected subclone).

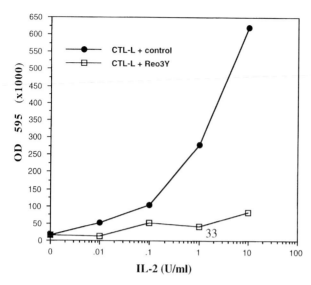

Fig. 2. Inhibition of proliferation by monoclonal antibody (mAb) 87.92.6 mimetics. Interleukin (*IL*)-2 dependent cytotoxic T lymphocyte (*CTL*)-L cells (5000 per well) were cultured in 96-well plates in media with 5% serum and increasing concentrations of IL-2. Peptides were added at 125 n*M*. Reo3Y is an analogue of mAb 87.92.6, and control was a random peptide with the same amino acid composition. After 24–36 h of culture, the tetrazolium salt (MTT) method was used to assess cell viability and proliferation as described (SARAGOVI et al. 1995). Optical densities (*OD*) were read at 595 nm wavelength

approximately 40% to 70%, as previously reported. In contrast, subclones expressing v-Ha-*ras* were not growth inhibited by p65/p95 binding (Table 2). Thus the antiproliferative effects of p65/p95 binding can be prevented by oncogenic p21^{ras} that is not under the control of GTPase-activating protein (GAP) (BOGUSKI and McCORMICK 1993; COOK and McCORMICK 1993).

The antimitotic effects caused by p65/p95-binding agents are significant even in the presence of serum (Table 2) or mitogens such as phorbol esters, calcium ionophores, or concanavilin A (SARAGOVI et al. 1991, 1995). More strikingly, in cultures of cell lines in which a defined growth factor drives proliferation (e.g., CTL-L cells dependent on IL-2), the antimitotic effect of p65/p95-binding agents is most evident and approaches 100% (Fig. 2). This effect is due to inhibition of IL-2 signals rather than inhibition of IL-2 binding to its receptor (data not shown).

9 Antiproliferative Activity In Vivo

We have performed in vivo experiments testing the immunosuppressive function of p65/p95 binding agents, particularly the small peptidomimetic Reo3Y. Allogeneic hearts transplanted under the ear pinna are fully rejected in approximately

12–15 days by immune-mediated mechanisms. The Reo3Y mimetic administered intravenously at doses of 100 μg per mouse prolonged the life of the grafted heart to longer than 25 days (C. MAIER, J. BROMBERG, and M.I. GREENE, unpublished results). In other studies, Reo3Y mimetics also significantly abolished the development of CD4$^+$ T cell-mediated immunity in a model of delayed-type hypersensitivity response (L. ZERBA and M.I. GREENE, unpublished results). Thus we conclude that the antimitotic/immunosuppressive effects mediated upon p65/p95 binding in vitro are significant and reproducible in vivo.

Other studies to date have indicated a proliferative effect when the Reo3Y mimetic or 87.92.6 antibody are applied to keratinocytes or corneal epithelial cell types (G. COTSARELLIS and M.I. GREENE, unpublished results). These data suggest that there may be additional receptor components or signaling mechanisms that differ in distinct tissues.

10 Putative Identity

The bioactivity of the p65/p95 ligands is reminiscent of other antiproliferative ligands such as transforming growth factor (TGF)-β. However, significant differences exist between TGF-β (MASSAGUÉ 1992a, b; LOPEZ-CASILLAS et al. 1993) and the p65/p95 complex. The main difference is that p95 appears to be a TK. Notably, Ser/Thr phosphorylation activity has also been detected for p95, and the activity may be more related to a dual-activity kinase family such as MEK (ZHENG and GUAN 1994) or GSK-3-β (WANG et al. 1994). Another important difference is that, while the antimitotic activity of TGF-β can be reduced by adding exogenous cytokines, this is not the case for p65/p95.

Several putative cellular receptors for HA3 have been reported by different laboratories. The two putative HA3 receptors most studied are (a) the β$_2$-adrenergic receptor (B2AR; LIU et al. 1988) or its family of proteins and (b) the EGFR (STRONG et al. 1993; TANG et al. 1993).

The B2AR has been reported to act as the HA3 receptor. However, by several criteria it cannot be a unique HA3 receptor. First, not all B2AR-expressing cells are bound by HA3 or mAb 87.92.6 (e.g., mouse OUB fibroblast cells transfected with B2AR cDNA; our unpublished data). Second, not all cell lines which are bound by mAb 87.92.6 or HA3 express B2AR (e.g., R1.1 thymoma). Third, catecholamine- and HA3-binding sites appear to be distinct and do not cross-compete with each other. Fourth, HA3, mAb 87.92.6, or reoviruses do not inhibit isoproterenol-induced cyclic adenosine monophosphate (cAMP) accumulation in DDT1 cells. Thus, if the B2AR is a reovirus-binding site, HA3 binds to a domain of the receptor distinct from the catecholamine-binding site (LIU et al. 1988).

Similarly, the EGFR cannot be the HA3 receptor, but it may be functionally linked to it. First, not all EGFR-expressing cells are bound by HA3 or mAb 87.92.6 (e.g., YAC cells). Second, some cell lines which do not express functional EGFR are bound by mAb 87.92.6 or HA3. However, it has been shown that reovirus binding is favored in cells expressing functional EGFR, but not in cells expressing a TK-dead mutant EGFR (STRONG et al. 1993).

Our interpretation of the data above is that, while EGFR does not directly bind HA3, it may be functionally associated with HA3 receptors. This would be particularly intriguing given our results that the p65/p95 receptor possesses activities reminiscent of EGFR, but of seemingly opposite action. There is a precedent in studies on *Drosophila* which identified a secreted natural protein related to EGF (the Argos protein) that inhibits EGFR activation. Argos mediates its action by binding to the extracellular domain of the EGFR (SCHWEITZER et al. 1995), and Argos is an inhibitor of both ligand-mediated and of ligand-independent activation.

11 Conclusions

We have shown that p65/p95 regulates cell cycle progression, DNA synthesis, and mitotic and perhaps meiotic cell division. It is suggested that p65/p95 regulation occurs via a novel mechanism that involves a cell surface TK. The biological relevance of p65/p95 receptors is emphasized by the dramatic effects in tyrosine phosphorylation and dephosphorylation of cellular substrates and ultimately in inhibition of mitosis, possibly through a $p21^{ras}$ pathway. The study, eventual cloning, and reconstitution of p65/p95 will provide valuable information concerning normal cell cycle control and its disregulation in neoplasias. Furthermore, the generation of tools and probes such as peptidomimetics may provide useful diagnostic or clinical agents in pathologies in which the complex or its signals are disregulated.

Acknowledgements. H.U.S. is a recipient of a Pharmaceutical Manufacturers Association of Canada-Medical Research Council (PMAC-MRC) Scholar Award. This work was supported by the Medical Research Council of Canada (H.U.S) and by the National Institutes of Health and the Lucille Markey Charitable Trust (M.I.G). We are grateful to Dr. Curtis Maier, Dr. Loukia Zerba, and Dr. George Cotsarellis (University of Pennsylvania) for sharing their unpublished data.

References

Blomquist MC, Hunt L, Barker W (1984) Vaccinia virus 19 kilodalton protein: relationship to several mammalian proteins, including two growth factors. Proc Natl Acad Sci USA 81:7363–7367
Boguski MS, McCormick F (1993) Proteins regulating ras and its relatives. Nature 366:643–653
Bruck C, Co MS, Slaoui M, Gaulton G, Smith T, Mullins J, Fields B, Greene MI (1986) Nucleic acid sequence of an internal image-bearing monoclonal anti-idiotype and its comparison to that of the external antigen. Proc Natl Acad Sci USA 83:6578–6582
Co M, Gaulton G, Fields B, Greene MI (1985a) Isolation and characterization of the reovirus hemagglutinin receptors. Proc Natl Acad Sci USA 82:1494–1498
Co M, Gaulton G, Fields B, Homcy M, Greene MI (1985b) Structural similarities of the reovirus and beta adrenergic receptors. Proc Natl Acad Sci USA 82:5315–5318

Cohen JA, Williams WV, Weiner DB, Greene MI (1990a) Antigenic and structural features of reoviruses. Immunochem Viruses 2:381–402

Cohen JA, Williams WV, Weiner DB, Geller HM, Greene MI (1990b) Ligand binding to the cell-surface receptor for reovirus type 3 stimulates galactocerebroside expression by developing oligodendrocytes. Proc Natl Acad Sci USA 87:4922–4926

Cook SJ, McCormick F (1993) Inhibition by cAMP of Ras-dependent activation of Raf. Science 262:1069–1072

Dichter MA, Weiner HL, Fields BN, Mitchell G, Noseworthy J, Gaulton GN, Greene MI (1986) Antiidiotypic antibody to reovirus binds to neurons and protects from viral infection. Ann Neurol 19:555–558

Dym M, Clermont Y (1970) Role of spermatogonia in the repair of the seminiferous epithelium following x-irradiation of the rat testis. Am J Anat 128:265–282

Epstein RL, Powers MK, Rogart RB, Weiner HL (1984) Binding of ^{125}I-labeled reovirus to cell surface receptors. Virology 133:46–55

Ertl H, Greene MI, Noseworthy J, Fields B, Nepom J, Spriggs D, Finberg E (1982) Identification of idiotypic receptors on reovirus specific cytolytic T cells. Proc Natl Acad Sci USA 79:7479–7483

Frade R, Barel M, Ehlin-Henriksson B, Klein G (1985) gp140, the c3d receptor of human B lymphocytes, is also the Epstein-Barr virus receptor. Proc Natl Acad Sci USA 82:1490–1493

Gaulton GN, Greene MI (1989) Inhibition of cellular DNA synthesis by reovirus occurs by a reovirus receptor linked signalling pathway which is mimicked by anti-receptor antibody. J Exp Med 169:197–212

Gentsch J, Pacitti A (1985) Effect of neuraminidase treatment of cells and effect of soluble glycoproteins on type 3 Reovirus attachment of murine L cells. J Virol 56:356

Greene MI, Kokai Y, Gaulton GN, Powell MB, Geller H, Cohen J (1987) Receptor systems in tissues of the nervous system. Immunol Rev 100:153–184

Kauffman R, Noseworthy J, Nepom J, Finberg R, Fields B, Greene MI (1983) Cell receptors for the mammalian reovirus. II. Monoclonal anti-idiotypic antibody blocks viral binding to cells. J Immunol 131:2539–2541

Kilham L, Margolis G (1969) Hydrocephalus in hamsters, ferrets, rats and mice following inoculations with reovirus type 1. II. Pathologic studies. Lab Invest 21:189–198

Klatzmann D, Champagne E, Chamaret S, Gruest J, Guetard D, Hercend T, Gluckman JC, Montagnier L (1984) T-lymphocyte T4 molecule behaves as the receptor for human retrovirus LAV. Nature 312:767–768

Lazaris-Karatzas A, Smith MR, Frederickson RM, Jaramillo ML, Liu Y-l, Kung H-F, Sonenberg N (1992) Ras-mediates translation initiation factor 4E-induced malignant transformation. Genes Dev 6(9):1631–1642

Lee PW, Hayes EC, Joklik WK (1981) Protein sigma 1 is the reovirus cell attachment protein. Virology 108:156–163

Li J-P, D'Andrea AD, Lodish HF, Baltimore D (1990) Activation of cell growth by binding of Friend-Spleen-focus forming virus gp55 glycoprotein to the erythropoietin receptor. Nature 343:762–764

Liu J, Co M, Greene MI (1988) Reovirus type 3 and [I125] iodocyanopindalol bind to distinct domains of the reovirus receptor. Immunol Res 7:232–238

Lopez-Casillas F, Wrana JL, Massague J (1993) Betaglycan presents ligand to the TGF beta signaling receptor. Cell 73:1435–1444

Masri SA, Nagata L, Mah DCW, Lee PWK (1986) Functional expression in Escherichia coli of cloned reovirus S1 gene encoding the viral cell attachment protein s-1. Virology 149:83–90

Massague J, Weinberg RA (1992a) Negative regulators of growth. Curr Opin Genet Dev 2:28–32

Massague J (1992b) Receptors for the TGF-beta family. Cell 69:1067–1070

Matsumoto AM (1996) Spermatogenesis. In: Adashi EY, Rock JA, Rosenwaks Z (eds) Reproductive endocrinology, surgery and technology, vol I. Lippincott-Raven, Philadelphia, pp 359–384

Matsuzaki N, Hinshaw V, Fields B, Greene MI (1986) Cell receptors for the mammalian reovirus: reovirus-specific T-cell hybridomas can become persistently infected and undergo autoimmune stimulation. J Virol 60:259

Nepom J, Weiner HL, Dichter MA, Fields BN, Greene MI (1982) Identification of a hemagglutinin specific idiotype associated with reovirus recognition shared by lymphoid and neuronal cells. J Exp Med 155:155–167

Noseworthy J, Fields B, Dichter M, Sobotka C, Pizer E, Perry L, Nepom J, Greene MI (1983) Cell receptors for the mammalian reovirus. I. Syngeneic monoclonal anti-idiotypic antibody identifies a cell surface receptor for reovirus. J Immunol 131:2533–2538

Pacitti A, Gentsch J (1987) Inhibition of reovirus type 3 binding to host cells by sialylated glycoproteins is mediated through the viral attachment protein. J Virol 61:1407–1415

Paul RW, Lee PKW (1987) Glycophorin is the reovirus receptor on human erythrocytes. Virology 159:94–101

Rebai N, Almazan G, Wei L, Greene M, Saragovi HU (1996) A p65/p95 neural receptor expressed at the S-G2 phase of the cell cycle subserves reovirus binding and defines distinct populations. Eur J Neurosci 8:273–281

Roux E, Robaire B, Saragovi HU (in press) A cell cycle regulating receptor is localized on the cell surface and in the nuclei of mitotically and meiotically dividing cells

Saragovi HU, Fitzpatrick D, Raktabutr A, Nakanishi H, Kahn M, Greene MI (1991) Design and synthesis of a mimetic of an antibody complementarity determining region. Science 253:792–795

Saragovi HU, Sauvé GJ, Greene MI (1992a) Viral receptors. In: Webster RG, Granoff A (eds) Encyclopedia of virology. Saunders, London

Saragovi HU, Chrusciel RA, Greene MI, Kahn M (1992b) Loops and secondary structure mimetics: development and applications in basic science and drug design. Biotechnology 10:773–778

Saragovi HU, Bhandoola A, Lemercier M, Akbar GKW, Greene MI (1995) A novel tyrosine kinase receptor subserves reovirus binding and regulates lymphocyte proliferation. DNA Cell Biol 14:653–664

Sauvé GJ, Saragovi HU, Greene MI (1992) Reovirus receptors. Adv Virus Res 42:325–341

Schweitzer R, Howes R, Smith R, Ben-Zion S, Freeman M (1995) Inhibition of Drosophila EGF receptor activation by the secreted protein Argos. Nature 376:699–701

Sharpe A, Gaulton G, McDade K, Fields B, Greene MI (1984) Syngeneic monoclonal anti-idiotype can induce cellular immunity to reovirus. J Exp Med 160:1195–1205

Siegel RM, Katsumata M, Komori S, Wadsworth S, Gill-Morse L, Jerrold-Jones S, Bhandoola A, Greene MI, Yui K (1990) Mechanisms of autoimmunity in the context of T cell tolerance. Immunol Rev 118:165–192

Strong JE, Tang D, Lee PWK (1993) Evidence that the epidermal growth factor receptor on host cells confers reovirus infection efficiency. Virology 197:405–411

Szyf M, Theberge J, Bozovic V (1995) Ras induces a general DNA methylation activity in mouse embryonal P19 cells. J Biol Chem 270:12690–12696

Tang D, Strong JE, Lee PWK (1993) Recognition of the epidermal growth factor receptor by reovirus. Virology 197:412–414

Tardieu M, Epstein R, Weiner H (1982) Interaction of viruses with cell surface receptors. Int Rev Cytol 80:27–61

Tyler KL, Fields BN (1990) Reoviridae and reoviruses. In: Second ED, Fields BN, Knipe DM et al. (eds) Virology. Raven, New York

Ullman KS, Northrop JP, Verweij CL, Crabtree G (1990) Transmission of signals from the T lymphocyte antigen receptor. The missing link. Annu Rev Immunol 8:421–452

Ullrich A, Schlessinger J (1990) Signal transduction by receptors with tyrosine kinase activity. Cell 61:203–212

Wang QM, Fiol CJ, DePaoli-Roach AA, Roach PJ (1994) Glycogen synthase kinase-3 beta is a dual specificity kinase differentially regulated by tyrosine and serine/threonine phosphorylation. J Biol Chem 269:14566–14574

Weiner HL, Ault KA, Fields BN (1980) Interaction of reovirus with cell surface receptors I: murine and human lymphocytes have a receptor for the hemagglutinin of reovirus type 3. J Immunol 124:2143–2148

Williams WV, Guy HR, Rubin DH, Robey F, Myers JN, Emmons TK, Weiner DB, Greene MI (1988) Sequences of the cell-attachment sites of reovirus type 3 and its anti-idiotypic/antireceptor antibody: modeling of their three-dimensional structures. Proc Natl Acad Sci USA 85:6488–6492

Williams WV, Moss DA, Kieber-Emmons T, Cohen J, Myers J, Weiner D, Greene MI (1989) Development of biologically active peptides based on antibody structure. Proc Natl Acad Sci USA 86:5537–5541

Zheng CF, Guan KL (1994) Activation of MEK family kinases requires phosphorylation of two conserved Ser/Thr residues. EMBO J 13:1123–1131

Reovirus Capsid Proteins σ3 and μ1: Interactions That Influence Viral Entry, Assembly, and Translational Control

L.A. Schiff

1 Introduction

Mammalian reoviruses are prototypical members of the *Reoviridae* family. Characteristics shared by members of this family include a segmented dsRNA genome and a nonenveloped particle consisting of two or three concentric proteinaceous capsids. The ten genome segments of mammalian reoviruses together encode only 11 unique proteins. These proteins execute the reovirus replication program in a variety of cell types within diverse animal hosts.

Reoviruses demonstrate considerable genetic economy compared to more complex RNA or DNA viruses that encode one or more dedicated regulatory proteins in addition to structural proteins and proteins with enzymatic functions. Eight of the 11 reovirus-encoded proteins are structural components of the virion; within the virion, there are several latent enzymatic activities which give rise to capped viral transcripts early in infection. Two of the three nonstructural reovirus proteins function as morphogenetic factors during the early stages of viral assembly. The function of the third nonstructural protein is not known. Interestingly,

Department of Microbiology, University of Minnesota, Minneapolis, MN 55455, USA

LIVERPOOL
JOHN MOORES UNIVERSITY
AVRIL ROBARTS LRC
TEL. 0151 231 4022

several recent studies suggest that some reovirus structural proteins also serve regulatory functions and affect the pathogenic process by influencing normal pathways involved in cellular viability and gene expression.

This chapter will focus on the structure and function of σ3 and μ1, the two most abundant proteins in the reovirus virion. During the process of entry into a susceptible cell, these capsid proteins undergo specific proteolytic cleavage events, one or more of which is critical to infection. Within infected cells, σ3 and μ1 are synthesized to high levels. In addition to serving a structural role in progeny virions, newly synthesized σ3 influences levels of cellular translation in infected cells. Protein μ1 also appears to have functions in the infected cell that are unrelated to its role in the capsid. Association of μ1 with σ3 abrogates the ability of σ3 to stimulate translation in transfected cells. Furthermore, although the mechanism responsible is not known, genetic evidence indicates that polymorphisms in capsid protein μ1 influence the ability of different reovirus strains to induce apoptosis (programmed cell death). Interestingly, the primary determinant of apoptosis in reovirus-infected cells is another structural protein: minor outer-capsid protein σ1. The functions of this protein are discussed in several other chapters in this volume. Following a brief description of reovirus virions and the structure of major capsid proteins σ3 and μ1, the remainder of the chapter will focus on how interactions between these two proteins influence their functions in the reovirus life cycle.

2 Structure of the Reovirus Virion

The eight reovirus structural proteins are arranged in two concentric icosahedral capsids (MAYOR et al. 1965; SMITH et al. 1969; AMANO et al. 1971; LUFTIG et al. 1972; Fig. 1, left). Four proteins comprise the inner capsid or core: σ2, λ1, λ3, and μ2. The arrangement of the inner-capsid subunits is unknown; however, biochemical studies suggest that proteins σ2 and λ1, which are present in large copy number, form the icosahedral lattice (WHITE and ZWEERINK 1976), whereas proteins λ3 and μ2, present in approximately 12 copies each, probably exist in a complex near the vertices. The reovirus outer capsid is also comprised of four proteins: λ2, σ1, σ3, and μ1. The 60 subunits of λ2 and the 600 subunits of μ1 form the T = 13 icosahedral lattice of the outer capsid (METCALF 1982; DRYDEN et al. 1993). The 36–48 copies of the cell attachment protein σ1 are found at the vertices (LEE et al. 1981a; FURLONG et al. 1988), and the 600 subunits of capsid protein σ3 occupy external positions on the outer capsid, in close apposition to capsid protein μ1 (DRYDEN et al. 1993).

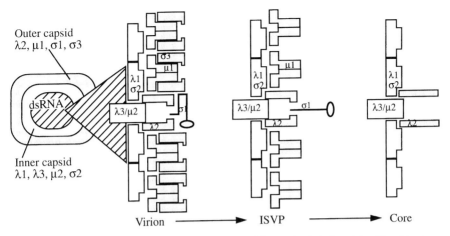

Fig. 1. Protein composition and organization of reovirus virions and subviral particles. The virion has a full complement of outer-capsid proteins. In the intermediate subviral particle (*ISVP*), there is loss of σ3, carboxy-terminal endoproteolytic cleavage of µl, and extension of σ1. In the core, there is loss of µl and σ1 and conformational change in λ2

3 Biochemical and Structural Analysis of Capsid Proteins σ3 and µl

The vigorous growth of most mammalian reovirus strains in cell culture and the ease with which reovirus particles can be purified from infected cells have facilitated biochemical and structural studies of the abundant capsid proteins. Recently, cloned cDNA copies of the S4 and M2 genes have been used for detailed molecular analysis of their gene products, σ3 and µl.

3.1 σ3 Protein

σ3 has two ligand-binding activities (Fig. 2) that can be assayed biochemically. Over 20 years ago, filter-binding assays using labeled, infected-cell extracts revealed that σ3 binds dsRNA in a sequence-independent manner (HUISMANS and JOKLIK 1976). More recently, ^{32}P-labeled dsRNA has been used to probe proteolytic fragments of σ3 and truncation mutants in northwestern blotting assays (SCHIFF et al. 1988; MILLER and SAMUEL 1992). These studies localized dsRNA-binding activity to a carboxy-terminal region of σ3 between residues 234 and 297. Within this region of σ3 lie two copies of a basic amino acid motif found in other dsRNA-binding proteins (MILLER and SAMUEL 1992). Alignment of these sequences and their predicted secondary structures suggests that the carboxy-terminal dsRNA-binding domain of σ3 has a similar three-dimensional structure to that of other dsRNA-binding proteins, including the interferon-induced dsRNA-dependent protein kinase PKR and the *Drosophila* staufen protein (YUE and SHATKIN 1996).

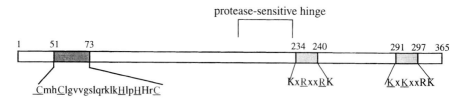

Fig. 2. Capsid protein σ3 (41 kDa), illustrating known or proposed ligand-binding sites. Within the carboxy-terminal RNA-dependent protein kinase (PKR)-like dsRNA-binding sites, basic residues are indicated in *capital letters*. Residues shown to be important in σ3's dsRNA-binding activity are *underlined*. Within the amino-terminal zinc-binding domain, potential ligand binding residues are indicated in *capital letters*. Residues shown to affect zinc-binding activity are *underlined*. Also indicated is the approximate location of a protease-hypersensitive region of σ3

Site-directed mutagenesis studies indicate that amino acids at positions 236, 239, 291, and 293 influence the dsRNA-binding activity of σ3 (DENZLER and JACOBS 1994). Strong evidence links this binding activity to σ3's ability to influence translation (IMANI and JACOBS 1988; GIANTINI and SHATKIN 1989; LLOYD and SHATKIN 1992; BEATTIE et al. 1995; MABROUK et al. 1995). This function of σ3 will be discussed in greater detail below in the section on translational control.

When the first predicted σ3 amino acid sequence was analyzed, an amino-terminal zinc-binding motif (between residues 51 and 73) was identified (SCHIFF et al. 1988). Atomic absorption analysis of virions, intermediate subviral particles (ISVP), and cores revealed that a single zinc atom was associated with each molecule of virion σ3; zinc was also stoichiometrically associated with σ3 purified from infected cells (SCHIFF et al. 1988). The observation that the zinc-binding and dsRNA-binding activities localize to distinct proteolytic fragments of σ3 indicated that σ3 does not conform to the zinc-binding protein paradigm established by transcription factor IIIA in which zinc-binding motifs contribute to a local structure that is directly involved in nucleic acid binding (SCHIFF et al. 1988). Site-directed mutagenesis studies suggest that the zinc finger contributes to the native structure and intracellular stability of σ3. When expressed transiently in transfected cells, σ3 zinc finger mutants show decreased intracellular stability (MABROUK and LEMAY 1994a). In vitro, σ3 zinc finger mutants fail to interact with μ1 and demonstrate increased sensitivity to proteolytic degradation (SHEPARD et al. 1996). Thus it is hypothesized that the zinc finger in σ3 is an important structural determinant that influences the ability of σ3 to form appropriate protein–protein interactions (SHEPARD et al. 1996).

3.2 μ1 Protein

Important biochemical features of the μ1 protein include its amino-terminal modification and proteolytic processing (Fig. 3). When reovirus virions are analyzed on sodium dodecyl sulfate (SDS)-polyacrylamide gels, the M2 gene product, μ1, appears largely in the form of two fragments, μ1N and μ1C (SMITH et al. 1969;

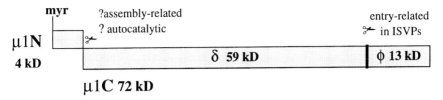

Fig. 3. Capsid protein μ1 (76 kDa), illustrating cleavage sites (*scissors*) and stable fragments μ1N, μ1C, δ, and φ. *ISVP*, intermediate subviral particle

ZWEERINK and JOKLIK 1970; CROSS and FIELDS 1976; LEE et al. 1981b). These fragments result from the endoproteolytic cleavage of μ1 between residues 42 and 43 (PETT et al. 1973; JAYASURIYA 1991). The amino-terminal fragment, μ1N, like the amino terminus of the precursor molecule, μ1, is modified by addition of the C_{14} fatty acid myristate (NIBERT et al. 1991). This modification is hypothesized to play a role in the ability of μ1/μ1N to interact with cell membranes during viral penetration into the cytoplasm. As discussed below, site-directed mutagenesis studies support the hypothesis that association with σ3 is required for the amino-terminal cleavage event that generates μ1N and μ1C (TILLOTSON and SHATKIN 1992); however, the importance of this cleavage for the reovirus life cycle is un-known.

During the early stages of infection in cell culture (CHANG and ZWEERINK 1971; SILVERSTEIN et al. 1972; BORSA et al. 1981; STURZENBECKER et al. 1987) and within the intestinal lumen following peroral inoculation (BODKIN et al. 1989; BASS et al. 1990), μ1 and μ1C undergo a carboxy-terminal endoproteolytic cleavage event. A similar cleavage occurs between residues 581 and 582 or between residues 584 and 585 when virions are digested in vitro with chymotrypsin or trypsin, respectively (NIBERT and FIELDS 1992). The carboxy-terminal fragment resulting from μ1 and μ1C cleavage has been referred to as φ and the amino-terminal fragments have been named μ1δ and δ, respectively. These fragments, like μ1N, remain stoichiometric-ally associated with the intermediate subviral particle (NIBERT et al. 1991; NIBERT and FIELDS 1992).

4 Functions of Outer-Capsid Proteins σ3 and μ1 During Viral Replication

4.1 σ3–μ1 Interactions That Influence Particle Stability

Proteins σ3 and μ1 interact extensively within the outer capsid. Sequences within σ3 and μ1 which are critical for their stable association have been identified using genetic and molecular approaches. These are discussed below in the context of viral assembly. Image reconstruction from cryoelectron micrographs of virions and subviral particles reveals that each σ3 subunit projects several nanometers above the surface of μ1 on the virion surface (DRYDEN et al. 1993). μ1 is organized in trimeric complexes that interact extensively at lower radii within the outer capsid (METCALF 1982; DRYDEN et al. 1993). Each σ3 molecule contacts as many as three adjacent μ1 molecules; some σ3 molecules in the outer capsid also contact protein λ2, which comprises the core spikes at the icosahedral vertices (DRYDEN et al. 1993). The extensive contacts between σ3 and μ1 molecules have been suggested to stabilize outer-capsid structure and thereby facilitate survival of virions during transmission of reovirus between hosts (NIBERT et al. 1996). Reovirus virions are remarkably stable in the environment and are relatively resistant to treatments with organic solvents, detergents, extreme pH values, and high temperature (SABIN 1959; GOMATOS et al. 1962; RHIM et al. 1961; WALLIS et al. 1964; DRAYNA and FIELDS 1982a, b). Outer-capsid proteins σ3, μ1, and σ1 have each been associated with the stability of reovirus particles against particular physicochemical agents (DRAYNA and FIELDS 1982a, b; WESSNER and FIELDS 1993; HOOPER and FIELDS 1996b).

4.2 Proteolysis of Capsid Proteins σ3 and μ1 in Early Steps of Reovirus Infection

Studies in vivo and in vitro reveal that susceptibility of reovirus outer-capsid proteins to proteolysis can critically influence infection. The first step in reovirus infection is attachment of the viral particle to receptor molecules on the target cell. Bound particles are subsequently internalized by receptor-mediated endocytosis (BORSA et al. 1979, 1981; STURZENBECKER et al. 1987; RUBIN et al. 1992). Binding is mediated by the cell attachment protein σ1, which extends from the vertices of the icosahedral viral particle (WEINER et al. 1978, 1980; LEE et al. 1981a). Polymorphisms in the cell attachment protein σ1 have been shown to affect conformation and susceptibility of this molecule to extracellular proteolysis (NIBERT et al. 1995); susceptibility of σ1 to intestinal proteases *negatively* affects pathogenesis in the host animal. Other studies suggest that extracellular proteolytic processing is actually *required* for reovirus to adhere to (AMERONGEN et al. 1994) and initiate productive infection in intestinal M cells (BASS et al. 1990).

The requirement for capsid proteolysis in vivo is beginning to be investigated at a molecular level. Following oral inoculation, virions are converted to ISVPs by the action of intestinal proteases (BODKIN et al. 1989; BASS et al. 1990). These ISVPs are virtually identical to particles generated in vitro by treatment of virions with chymotrypsin (BODKIN et al. 1989). Biochemical and structural analysis of in vitro-generated ISVPs (Fig. 1, middle) reveals that they have an extended form of the cell attachment protein σ1 (FURLONG et al. 1988), lack outer-capsid protein σ3, and have μ1 molecules which have been endoproteolytically cleaved near their carboxy termini (SMITH et al. 1969; JOKLIK 1972; BORSA et al. 1973). It has been suggested that intraluminal proteolysis may be required for viral cell attachment protein σ1 to undergo the conformational change that allows it to contact M cell apical receptors (NEUTRA et al. 1996).

Reovirus infection of cultured cells and extraintestinal targets (including endothelial cells of the nervous system and M cells within the respiratory tract; MORIN et al. 1994) is likely initiated by intact virions that require intracellular proteolysis. It is believed that the acid-dependent proteolytic digestion of intact reovirions is an essential step in the replication cycle in mouse L929 fibroblasts, since treatment of cells with agents that raise intracellular pH and block proteolysis effectively inhibits replication (STURZENBECKER et al. 1987). This block can be overcome by infecting cells with ISVPs prepared by in vitro proteolysis (MARATOS-FLIER et al. 1986; STURZENBECKER et al. 1987; DERMODY et al. 1993). Preliminary studies with protease inhibitors suggest that a cellular, E64-sensitive lysosomal cysteine protease is involved in reovirus uncoating (M.L. NIBERT, personal communication).

In vitro-generated ISVPs have unique biological activities that suggest that their in vivo-generated correlates mediate penetration from the endosomal compartment into the cytoplasm. These ISVPs can perturb cellular membranes in assays designed to measure ^{51}Cr release or conductance through artificial planar bilayers, whereas intact virions and core particles are not functional in these assays (LUCIA-JANDRIS et al. 1993; TOSTESON et al. 1993). Cores (Fig. 1, right) are a distinct type of subviral particle that lack all outer-capsid proteins. These results suggest that ISVP-like particles generated in the intestinal lumen may be capable of penetrating intestinal M cells directly at the plasma membrane.

Current efforts are aimed at identifying which of the structural changes in ISVPs are critical to their ability to interact with membranes and bypass the block to replication imposed by lysosomotropic agents such as ammonium chloride. The only loss of mass associated with the conversion of virions to ISVPs is a consequence of proteolytic removal of σ3. μ1 undergoes only minor conformational changes during this process, despite its endoproteolytic cleavage (DRYDEN et al. 1993). It seems likely that determinants within μ1 are directly involved in membrane penetration, since a strain difference in the capability of ISVPs to cause ^{51}Cr release was mapped to the M2 gene that encodes μ1 (LUCIA-JANDRIS et al. 1993). Other studies of M2 gene mutants and protective μ1-specific monoclonal antibodies point towards the importance of a central region of μ1 in this process (HAZELTON and COOMBS 1995; HOOPER and FIELDS 1996a, b). However, these facts do not preclude the possibility that the pH-dependent proteolytic removal of σ3 is required to reveal polymorphic membrane-interactive sequences that are present in both cleaved and uncleaved forms of μ1. Additional experiments are required to determine whether it is proteolytic removal of σ3, the carboxy-terminal cleavage of μ1/μ1C, or both that renders ISVPs membrane interactive.

Proteolytic uncoating eventually yields core particles which lack all outer-capsid proteins. In vitro-generated cores are transcriptionally active (BORSA and GRAHAM 1968; BANERJEE and SHATKIN 1970), and their in vivo counterparts are presumed to be responsible for primary viral transcription (SILVERSTEIN and DALES 1968; CHANG and ZWEERINK 1971; SILVERSTEIN et al. 1972; BORSA et al. 1981). It remains unclear whether reovirus transcription is regulated. There is some evidence that four reovirus mRNAs (S4, L1, S3, and M3) are expressed earlier than the remaining six (WATANABE et al. 1968; NONOYAMA et al. 1974; SPANDIDOS et al.

1976) and that transcription from the remaining genes requires new protein synthesis (WATANABE et al. 1968; SHATKIN and LaFIANDRA 1972; LAU et al. 1975). These data suggest that one or more proteins encoded by the pre-early mRNAs (σ3, σNS, μNS, and λ3) are required for full genomic transcription.

4.3 Role of σ3–μ1 Interactions in Translational Control

Synthesis of reoviral proteins from capped viral transcripts is mediated by the host cell translational machinery. Following infection with most reovirus strains, viral protein synthesis occurs efficiently despite a dramatic inhibition of translation of cellular proteins (GOMATOS and TAMM 1963; ZWEERINK and JOKLIK 1970; SHARPE and FIELDS 1982). As in many viral systems, the mechanism by which viral protein synthesis comes to predominate in reovirus infected cells is not well understood. Some data in the literature suggest that reovirus mRNAs are preferentially translated in infected cells because reovirus infection results in a modification of the cap dependence of the host cell translational machinery (SKUP and MILLWARD 1980b) such that uncapped late viral mRNAs are preferentially translated (SKUP and MILLWARD 1980a; ZARBL et al. 1980; LEMIEUX et al. 1984). Other studies suggest that viral mRNAs are preferentially translated because they are abundant and compete with cellular mRNAs for limiting translation factors (WALDEN et al. 1981; RAY et al. 1983). The results of recent coinfection experiments conflict with both of these proposed mechanisms (SCHMECHEL et al. 1997). Cellular translation is spared in L cells coinfected with the Jones strain, which is strongly inhibitory to cellular translation, and the Dearing strain, which spares cellular translation. Similar results were obtained using the inhibitory Abney strain (c87). If efficient viral translation requires inactivation of a factor required for cap-dependent translation of cellular RNA, inhibition of cellular translation would be expected to be *dominant* in co-infected cells. The observation that translational sparing is dominant in cells co-infected with inhibitory and noninhibitory strains is also incompatible with a model in which inhibitory strains synthesize abundant viral mRNAs which compete with cellular mRNAs for limiting translational factors. Quantitative analysis of viral and cellular mRNAs in infected cells similarly fails to support the hypothesis that the ratio of viral to cellular mRNA influences the degree of inhibition of cellular translation (SCHMECHEL et al. 1997).

Several lines of evidence support the hypothesis that, in the infected cell, newly synthesized σ3 influences the balance of translation. Reassortant virus analysis mapped the polymorphism between two reovirus strains that differ in their degree of inhibition of cellular protein synthesis to the reovirus S4 gene encoding σ3 (SHARPE and FIELDS 1982). As mentioned above, σ3 binds dsRNA in a sequence-independent manner (HUISMANS and JOKLIK 1976). This activity could allow σ3 to block activation of the dsRNA-activated kinase PKR, an interferon-induced enzyme that inhibits translation initiation by phosphorylating eIF-2α (HOVANESSIAN 1991). In reovirus-infected cells, the S1 gene mRNA is believed to be a potent activator of PKR (SAMUEL and BRODY 1990; BELLI and SAMUEL 1993; HENRY et al. 1994).

In vitro (IMANI and JACOBS 1988) and in heterologous virus systems (LLOYD and SHATKIN 1992; BEATTIE et al. 1995), σ3 can inhibit the activation of PKR. σ3 can rescue adenovirus and vaccinia virus mutants with deletions (in VA1 and E3L, respectively) that render these viruses sensitive to the inhibitory effects of interferon. Residues that are important for σ3–dsRNA binding are required for rescue of E3L mutants (BEATTIE et al. 1995). This latter finding strongly suggests that, in vaccinia virus infected cells, σ3 functions to sequester the dsRNA activator of PKR, thus enabling translation of viral mRNA and efficient vaccinia virus replication.

Despite compelling evidence that σ3 can inhibit PKR and that polymorphisms in σ3 influence translation in reovirus-infected cells, the molecular basis for these polymorphisms is not well understood. Comparative sequence analysis revealed that the carboxy-terminal PKR-like dsRNA-binding sites are highly conserved among σ3 sequences from virus strains with distinct protein synthesis phenotypes (KEDL et al. 1995). The σ3 molecules from Dearing and Abney strains are entirely conserved within the dsRNA-binding motifs, yet these strains have dramatically different consequences on cellular translation. These results suggest either that inherent differences in dsRNA binding do not contribute to σ3-mediated differences in the level of cellular translation or that dsRNA binding is influenced by sequences outside of the well-characterized motifs.

A recent study of σ3 subcellular localization may provide insight into the molecular basis for σ3-related polymorphisms in cellular protein synthesis and the mechanism which enables viral translation to proceed efficiently, even in the face of dramatic inhibition of cellular translation (SCHMECHEL et al. 1997). Immunofluorescence analysis reveals that, in cells infected with the inhibitory Abney and Jones strains, σ3 is largely restricted to viral factories, whereas in cells infected with the sparing Dearing strain, σ3 is present diffusely throughout the cytoplasm and within the nucleus. These findings are consistent with a model (Fig. 4) in which σ3 *locally* blocks activation of PKR. PKR is known to be localized throughout the cytoplasm and also within the nucleus of infected cells (JIMENEZ-GARCIA et al. 1993; JEFFREY et al. 1995). Thus, in cells infected with most reovirus strains, PKR is activated throughout the cytoplasm to inhibit cellular translation, yet viral translation is spared by the local dsRNA-sequestering activity of σ3 within viral factories. Consistent with this model is the finding that eIF-2α is less phosphorylated in cells infected with the sparing Dearing strain than in cells infected with the inhibitory Jones strain (LLOYD and SHATKIN 1992). Also consistent is the observation that interferon-induced enzymes are upregulated in cells infected with the Dearing strain, in which viral and cellular translation proceed efficiently (FEDUCHI et al. 1988).

The molecular basis for the strain-specific differences in σ3 localization may involve the affinity of σ3 for outer-capsid protein μ1. These capsid proteins are found in a 1:1 ratio within the viral outer capsid and can be recovered in stable complexes from infected cells by immunoprecipitation with σ3- or μ1-specific monoclonal antibodies (SMITH et al. 1969; LEE et al. 1981b; JAYASURIYA 1991).

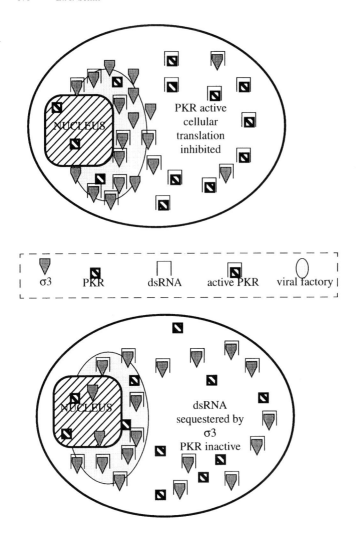

Fig. 4a, b. Model for spared viral translation in reovirus-infected cells. **a** Cell infected with reovirus strain that is inhibitory to cellular translation. RNA-dependent protein kinase (*PKR*) is inactive in the viral factory due to dsRNA sequestering activity of σ3, but is active in the rest of the cytoplasm. **b** Cell infected with reovirus strain that is sparing to cellular translation. PKR is inactive in both the viral factory and the rest of the cytoplasm due to dsRNA sequestering activity of σ3

Whereas the pattern of σ3 localization varies with the infecting strain, capsid protein μ1 appears to be restricted to viral factories in cells infected with all strains (SCHMECHEL et al. 1997). Immunoprecipitation studies reveal that differences in σ3 localization correlate with differences in σ3–μ1 affinity (SCHMECHEL et al. 1997). Dearing strain σ3, which has a diffuse cytoplasmic localization, is poorly co-immunoprecipitated using antibodies directed against μ1. In contrast, σ3 from the inhibitory Jones and Abney strains is efficiently coprecipitated with μ1 and is preferentially localized to perinuclear viral factories.

The hypothesis that efficient viral translation is a consequence of σ3-mediated local blockade of PKR activation requires that not all of the σ3 within viral factories is complexed with μ1. This is a critical feature of the model, since several pieces of data suggest that σ3–μ1 complexes are incapable of binding dsRNA. First, in transfection studies, coexpression of σ3 and μ1 has been found to abrogate the stimulatory effect of σ3 on reporter gene expression (TILLOTSON and SHATKIN 1992). Only those μ1 molecules which can form complexes with σ3 are active in this assay. Furthermore, σ3–μ1 complexes are not observed when σ3 is affinity purified on columns containing dsRNA (HUISMANS and JOKLIK 1976). In support of this model, preliminary confocal analysis suggests that not all σ3 in the viral factories is colocalized with μ1, and immunoprecipitation studies indicate that not all σ3 is complexed with μ1, even in cells infected with inhibitory strains.

4.4 σ3–μ1 Interactions in Virion Morphogenesis

Relatively little is known about the protein–protein and protein–RNA interactions that are required to assemble an infectious reovirus particle. Available evidence suggests that reovirus morphogenesis occurs in defined stages and that capsid protein σ3 may function at more than one stage. The first step in assembly has been suggested to be formation of ribonucleoprotein "assortment" complexes which contain viral mRNA, the nonstructural proteins μNS and σNS, and capsid protein σ3 (ANTCZAK and JOKLIK 1992). At this time, it is not clear whether the σ3 in these RNP complexes interacts with viral transcripts or one or both of the nonstructural proteins. These nascent progeny particles undergo structural rearrangements, losing nonstructural proteins and gaining core proteins. Within newly formed "replicase particles," the packaged mRNAs serve as templates for the synthesis of genomic minus strands (WARD et al. 1972; WARD and SHATKIN 1972; ZWEERINK et al. 1972; ZWEERINK 1974; MORGAN and ZWEERINK 1975). Some evidence suggests that a final step in viral assembly is the addition of preformed complexes of outer-capsid proteins σ3 and μ1 onto core-like replicase particles (FIELDS et al. 1971; MORGAN and ZWEERINK 1974).

Molecular determinants which influence the formation of σ3–μ1 complexes have been identified in studies with σ3 fragments, temperature-sensitive (ts) mutants, and site-directed point mutants (MABROUK and LEMAY 1994a; SHEPARD et al. 1996). Several of these studies suggest that the amino-terminal domain of σ3 contains determinants which influence μ1 binding. Substitution of serine for cysteine residues within the amino-terminal zinc finger results in mutant σ3 molecules that do not form stable complexes with μ1 (SHEPARD et al. 1996). As mentioned earlier, this is likely because the zinc finger contributes to the native conformation of the amino terminus of σ3. tsG mutants, which have alterations in the σ3 protein, have defects in virion morphogenesis at the restrictive temperature; the altered σ3 proteins are compromised in their ability to assemble σ3–μ1 complexes onto progeny core-like particles (FIELDS et al. 1971; MORGAN and ZWEERINK 1974). Sequence analysis of the prototypic *tsG453* mutant revealed six amino acid substitutions relative to the wild-type (wt) σ3 (at positions 16, 117, 138, 141, and 229). The proximity of the change at position 16 to the zinc finger and the frequency with which it was observed in independent clones led investigators to suggest that this change played a major role in the *ts* phenotype (DANIS et al. 1992). Other evidence suggests that σ3–μ1 interactions may be more extensive. Sequence analysis of other mutants suggests that amino acids within the first dsRNA-binding region also influence the σ3–μ1 interaction (KEDL and SCHIFF).

The formation of σ3–μ1 complexes results in structural rearrangements in both of these proteins. One important consequence of the association of σ3 and μ1 during viral assembly is to render σ3 protease sensitive (SHEPARD et al. 1995). As mentioned above, early in infection, virions are converted to ISVPs by the proteolytic removal of σ3 and the carboxy-terminal cleavage of μ1/μ1C. One or both of these cleavage events is required for viral infection to proceed. Molecular studies using in vitro-translated protein reveal that newly translated σ3 undergoes a conformational change upon association with μ1 that converts it from a protease-resistant form to a protease-sensitive form (SHEPARD et al. 1995). The reactivity of both of these σ3 forms with conformation-dependent monoclonal antibodies suggests that the μ1-mediated change does not dramatically alter the tertiary structure of σ3. Rather, association with μ1 causes a conformational shift in σ3 structure to expose a previously described protease-sensitive hinge region (SCHIFF et al. 1988; MILLER and SAMUEL 1992; DRYDEN et al. 1993) that is inaccessible in newly synthesized σ3.

A second assembly-related consequence of the σ3–μ1 interaction may be the amino-terminal cleavage of μ1. Immunoprecipitation studies reveal that, within the cytoplasm of reovirus-infected cells, 95% of the μ1-related protein that is complexed with σ3 is in the form of μ1C (LEE et al. 1981b). Studies in which wt and mutant S4 and M2 genes were cotransfected into COS cells support a role for σ3 in μ1 cleavage (TILLOTSON and SHATKIN 1992). Cells transfected with the M2 gene alone expressed intact μ1, whereas cells transfected with the S4 and M2 genes expressed σ3 and μ1C. Mutational analysis of μ1 reveals that the amino-terminal cleavage of μ1 is sensitive not only to residues at the cleavage junction, but also to alterations which affect the amino-terminal myristoylation. Mutant μ1 molecules which are not myristoylated also fail to form stable complexes with σ3, further supporting the hypothesis that σ3–μ1 association is a prerequisite for μ1 cleavage. Neither the importance of this amino-terminal cleavage for assembly or entry nor the role of σ3 are understood. Not all μ1 in virions has undergone this cleavage, since some intact μ1 is routinely observed in SDS-polyacrylamide gel electrophoresis (PAGE) analysis of purified virions.

Although a region of σ3 with limited sequence similarity to picornaviral proteases was identified (SCHIFF et al. 1988), there is no convincing evidence that σ3 is the protease that effects the amino-terminal μ1 cleavage (MABROUK and LEMAY 1994b). An alternative hypothesis for the effect of σ3 on μ1 cleavage is that association of σ3 with μ1 allows μ1 to assume a conformation that can undergo autocatalytic cleavage. This hypothesis is suggested by the sequence and structural similarities between the μ1 and the picornavirus capsid protein VP0. The cleavage of VP0 into capsid proteins VP2 and VP4 is autocatalytic, both VP0 and μ1 are myristoylated at their amino termini, and they share sequence similarity around their cleavage sites (NIBERT et al. 1991).

5 Concluding Remarks

Capsid proteins σ3 and μ1 are present in large numbers in the reovirus virion and in great abundance within infected cells. They function at a number of distinct steps in the viral life cycle, contributing to the genetic economy of reoviruses. They assume structural roles in viral entry and assembly, but also influence the virus host cell interaction in a qualitative way by affecting the balance of translation and by modulating the cellular apoptotic response. Interestingly, each of these proteins appears to influence the structure and function of the other. Molecular studies suggest that σ3 must form complexes with μ1 in order to assume the protease-sensitive conformation found in virions. This change is functionally significant, since proteolysis of σ3 is a critical early event in reovirus infection. Some evidence suggests that σ3 may be expressed somewhat earlier than μ1 in infected cells. Newly synthesized σ3 is relatively protease resistant, binds dsRNA, and can inhibit the activation of PKR. Later in infection, association of μ1 with σ3 may serve to modulate the translation-regulatory function of σ3. The formation of σ3–μ1 complexes may block σ3 from interacting with dsRNA and render it protease sensitive, thus shifting it from a translation-regulatory molecule to a functionally competent structural component. Assembly of μ1 onto progeny core particles may depend upon the formation of σ3–μ1 complexes within the cytoplasm, since *ts* mutants with alterations in σ3 assemble particles which lack both σ3 and μ1 at the nonpermissive temperature. The formation of σ3–μ1 complexes also appears to initiate the amino-terminal cleavage of μ1 into μ1N and μ1C through a mechanism that is not understood, but might involve conformational priming of μ1 autocatalysis. Future studies of reovirus protein structure and function will undoubtedly reveal other examples of proteins, similar to σ3 and μ1, whose interaction modulates and extends protein function.

Acknowledgements. The author would like to acknowledge the contribution of Dr. Bernard Fields, who championed the reovirus system and enthusiastically supported the development of young scientists. Stephen Schmechel provided valuable editorial advice during the preparation of this manuscript. Work in the author's laboratory was supported by a grant from the National Institutes of Health (AI 32139).

References

Amano Y, Katagiri S, Ishida N, Watanabe Y (1971) Spontaneous degradation of reovirus capsid into subunits. J Virol 8(5):805–808

Amerongen HM, Wilson GA, Fields BN, Neutra MR (1994) Proteolytic processing of reovirus is required for adherence to intestinal M cells. J Virol 68(12):8428–8432

Antczak JB, Joklik WK (1992) Reovirus genome segment assortment into progeny genomes studied by the use of monoclonal antibodies directed against reovirus proteins. Virology 187(2):760–776

Banerjee AK, Shatkin AJ (1970) Transcription in vitro by reovirus-associated ribonucleic acid-dependent polymerase. J Virol 6(1):1–11

Bass DM, Bodkin D, Dambrauskas R, Trier JS, Fields BN, Wolf JL (1990) Intraluminal proteolytic activation plays an important role in replication of type 1 reovirus in the intestines of neonatal mice. J Virol 64(4):1830–1833

Beattie EK, Denzler L, Tartaglia J, Perkus ME, Paoletti E, Jacobs BL (1995) Reversal of the interferon-sensitive phenotype of a vaccinia virus lacking E3L by expression of the reovirus S4 gene. J Virol 69(1):499–505

Belli BA, Samuel CE (1993) Biosynthesis of reovirus-specified polypeptides: identification of regions of the bicistronic reovirus S1 mRNA that affect the efficiency of translation in animal cells. Virology 193(1):16–27

Bodkin DK, Nibert ML, Fields BN (1989) Proteolytic digestion of reovirus in the intestinal lumens of neonatal mice. J Virol 63(11):4676–4681

Borsa J, Graham AF (1968) Reovirus: RNA polymerase activity in purified virions. Biochem Biophys Res Commun 33(6):895–901

Borsa J, Copps TP, Sargent MD, Long DG, Chapman JD (1973) New intermediate subviral particles in the in vitro uncoating of reovirus virions by chymotrypsin. J Virol 11(4):552–564

Borsa J, Morash BD, Sargent MD, Copps TP, Lievaart PA, Szekely JG (1979) Two modes of entry of reovirus particles into L cells. J Gen Virol 45(1):161–170

Borsa J, Sargent MD, Lievaart PA, Copps TP (1981) Reovirus: evidence for a second step in the intracellular uncoating and transcriptase activation process. Virology 111(1):191–200

Chang CT, Zweerink HJ (1971) Fate of parental reovirus in infected cell. Virology 46(3):544–555

Cross RK, Fields BN (1976) Reovirus-specific polypeptides: analysis using discontinuous gel electrophoresis. J Virol 19(1):162–173

Danis C, Garzon S, Lemay G (1992) Further characterization of the ts453 mutant of mammalian orthoreovirus serotype 3 and nucleotide sequence of the mutated S4 gene. Virology 190:494–498

Denzler KL, Jacobs BL (1994) Site-directed mutagenic analysis of reovirus sigma 3 protein binding to dsRNA. Virology 204(1):190–199

Dermody TS, Nibert ML, Wetzel JD, Tong X, Fields BN (1993) Cells and viruses with mutations affecting viral entry are selected during persistent infections of L cells with mammalian reoviruses. J Virol 76:2055–2063

Drayna D, Fields BN (1982a) Biochemical studies on the mechanism of chemical and physical inactivation of reovirus. J Gen Virol 63(1):161–170

Drayna D, Fields BN (1982b) Genetic studies on the mechanism of chemical and physical inactivation of reovirus. J Gen Virol 63:149–160

Dryden KA, Wang G, Yeager M, Nibert ML, Coombs KM, Furlong DB, Fields BN, Baker TS (1993) Early steps in reovirus infection are associated with dramatic changes in supramolecular structure and protein conformation: analysis of virions and subviral particles by cryoelectron microscopy and image reconstruction. J Cell Biol 122(5):1023–1041

Feduchi E, Esteban M, Carrasco L (1988) Reovirus type 3 synthesizes proteins in interferon-treated HeLa cells without reversing the antiviral state. Virology 164:420–426

Fields BN, Raine CS, Baum SG (1971) Temperature-sensitive mutants of reovirus type 3: defects in viral maturation as studied by immunofluorescence and electron microscopy. Virology 43:569–578

Furlong DB, Nibert ML, Fields BN (1988) Sigma 1 protein of mammalian reoviruses extends from the surfaces of viral particles. J Virol 62:246–256

Giantini M, Shatkin AJ (1989) Stimulation of chloramphenicol acetyltransferase mRNA translation by reovirus capsid polypeptide sigma 3 in cotransfected COS cells. J Virol 63(6):2415–2421

Gomatos PJ, Tamm I (1963) Macromolecular synthesis in reovirus-infected L cells. Biochim Biophys Acta 72:651–653

Gomatos PJ, Tamm I, Dales S, Franklin RM (1962) Reovirus type 3: physical characteristics and interactions with L cells. Virology 17:441–454

Hazelton PR, Coombs KM (1995) The reovirus mutant tsA279 has temperature-sensitive lesions in the M2 and L2 genes: the M2 gene is associated with decreased viral protein production and blockade in transmembrane transport. Virology 207(1):46–58

Henry GL, McCormack SJ, Thomis DC, Samuel CE (1994) Mechanism of interferon action. Translational control and the RNA-dependent protein kinase (PKR): antagonists of PKR enhance the translational activity of mRNAs that include a 161 nucleotide region from reovirus S1 mRNA. J Biol Regul Homeost Agents 8(1):15–24

Hooper JW, Fields BN (1996a) Monoclonal antibodies to reovirus sigma 1 and mu 1 proteins inhibit chromium release from mouse L cells. J Virol 70(1):672–677

Hooper JW, Fields BN (1996b) Role of the mu 1 protein in reovirus stability and capacity to cause chromium release from host cells. J Virol 70(1):459–467

Hovanessian AG (1991) Interferon-induced and double-stranded RNA-activated enzymes: a specific protein kinase and 2′,5′-oligoadenylate synthetases. J Interferon Res 11(4):199–205

Huismans H, Joklik WK (1976) Reovirus-coded polypeptides in infected cells: isolation of two native monomeric polypeptides with affinity for single-stranded and double-stranded RNA, respectively. Virology 70(2):411–424

Imani F, Jacobs BL (1988) Inhibitory activity for the interferon-induced protein kinase is associated with the reovirus serotype 1 sigma 3 protein. Proc Natl Acad Sci USA 85(21):7887–7891

Jayasuriya AKU (1991) Molecular characterization of the reovirus M2 gene. Microbiology and molecular genetics. Thesis, Harvard University

Jeffrey IW, Kadereit S, Meurs EF, Metzger T, Bachmann M, Schwemmle M, Hovanessian AG, Clemens MJ (1995) Nuclear localization of the interferon-inducible protein kinase PKR in human cells and transfected mouse cells. Exp Cell Res 218(1):17–27

Jimenez-Garcia LF, Green SR, Matthews MB, Spector DL (1993) Organization of the double-stranded RNA-activated protein kinase DAI and virus-associated VA RNAI in adenovirus-2-infected HeLa cells. J Cell Sci 106:11–22

Joklik WK (1972) Studies on the effect of chymotrypsin on reovirions. Virology 49:700–801

Kedl R, Schmechel S, Schiff L (1995) Comparative sequence analysis of the reovirus S4 genes from 13 serotype 1 and serotype 3 field isolates. J Virol 69(1):552–559

Lau RY, Van Alstyne D, Berckmans R, Graham AF (1975) Synthesis of reovirus-specific polypeptides in cells pretreated with cycloheximide. J Virol 16(3):470–478

Lee PWK, Hayes EC, Joklik WK (1981b) Characterization of anti-reovirus immunoglobulins secreted by cloned hybridoma cell lines. Virology 108:134–146

Lee W, Hayes EC, Joklik WK (1981a) Protein sigma 1 is the reovirus cell attachment protein. Virology 108(1):156–163

Lemieux R, Zarbl H, Millward S (1984) mRNA discrimination in extracts from uninfected and reovirus-infected L-cells. J Virol 51(1):215–222

Lloyd RM, Shatkin AJ (1992) Translational stimulation by reovirus polypeptide sigma 3: substitution for VAI RNA and inhibition of phosphorylation of the alpha subunit of eukaryotic initiation factor 2. J Virol 66(12):6878–6884

Lucia-Jandris P, Hooper JW, Fields BN (1993) Reovirus M2 gene is associated with chromium release from mouse L cells. J Virol 67(9):5339–5345

Luftig RB, Kilham SS, Hay AJ, Zweerink HJ, Joklik WK (1972) An ultrastructural study of virions and cores of reovirus type 3. Virology 48(1):170–181

Mabrouk T, Lemay G (1994a) Mutations in a CCHC zinc-binding motif of the reovirus sigma 3 protein decrease its intracellular stability. J Virol 68(8):5287–5290

Mabrouk T, Lemay G (1994b) The sequence similarity of reovirus sigma 3 protein to picornaviral proteases is unrelated to its role in mu 1 viral protein cleavage. Virology 202(2):615–620

Mabrouk T, Danis C, Lemay G (1995) Two basic motifs of reovirus sigma 3 protein are involved in double-stranded RNA binding. Biochem Cell Biol 73(3/4):137–145

Maratos-Flier E, Goodman MJ, Murray AH, Kahn CR (1986) Ammonium inhibits processing and cytotoxicity of reovirus, a nonenveloped virus. J Clin Invest 78(4):1003–1007

Mayor HD, Jamison RM, Jordan LE, Mitchell MV (1965) Reoviruses. II. Structure and composition of the virion. J Bacteriol 89:1548–1556

Metcalf P (1982) The symmetry of the reovirus outer shell. J Ultrastruct Res 78:292–301

Miller JE, Samuel CE (1992) Proteolytic cleavage of the reovirus sigma 3 protein results in enhanced double-stranded RNA-binding activity: identification of a repeated basic amino acid motif within the C-terminal binding region. J Virol 66(9):5347–5356

Morgan EM, Zweerink HJ (1974) Reovirus morphogenesis: core-like particles in cells infected at 39 with wild-type reovirus and temperature-sensitive mutants of groups B and G. Virology 59:556–565

Morgan EM, Zweerink HJ (1975) Characterization of transcriptase and replicase particles isolated from reovirus-infected cells. Virology 68(2):455–466

Morin MJ, Warner A, Fields BN (1994) A pathway for entry of reoviruses into the host through M cells of the respiratory tract. J Exp Med 180(4):1523–1527

Neutra MR, Frey A, Kraehenbuhl J-P (1996) Epithelial M cells: gateways for mucosal infection and immunization. Cell 86:345–348

Nibert ML, Fields BN (1992) A carboxy-terminal fragment of protein mu 1/mu 1C is present in infectious subvirion particles of mammalian reoviruses and is proposed to have a role in penetration. J Virol 66(11):6408–6418

Nibert ML, Schiff LA, Fields BN (1991) Mammalian reoviruses contain a myristoylated structural protein. J Virol 65:1960–1967

Nibert ML, Chappell JD, Dermody TS (1995) Infectious subvirion particles of reovirus type 3 Dearing exhibit a loss in infectivity and contain a cleaved sigma 1 protein. J Virol 69(8):5057–5067

Nibert ML, Schiff L, Fields BN (1996) Reoviruses and their replication. In: Fields BN, Knipe DM, Howley Fields PM (eds) Virology. Lippincott-Raven, Philadelphia

Nonoyama M, Millward S, Graham AF (1974) Control of transcription of the reovirus genome. Nucleic Acids Res 1:373–385

Pett DM, Vanaman TC, Joklik WK (1973) Studies on the amino and carboxyl terminal amino acid sequences of reovirus capsid polypeptides. Virology 52(1):174–186

Ray BK, Brendler TG, Adya S, Daniels-McQueen S, Miller JK, Hershey JW, Grifo JA, Merrick WC, Thach RE (1983) Role of mRNA competition in regulating translation: further characterization of mRNA discriminatory initiation factors. Proc Natl Acad Sci USA 80(3):663–667

Rhim JS, Smith KO, Melnick JL (1961) Complete and coreless forms of reovirus (ECHO 10): ratio of number of virus particles to infective units in the one-step growth cycle. Virology 15:428–435

Rubin DH, Weiner DB, Dworkin C, Greene MI, Maul GG, Williams WV (1992) Receptor utilization by reovirus type 3: distinct binding sites on thymoma and fibroblast cell lines result in differential compartmentalization of virions. Microb Pathog 12:351–365

Sabin AB (1959) Reoviruses. Science 130:1387–1389

Samuel CE, Brody MS (1990) Biosynthesis of reovirus-specified polypeptides: 2-aminopurine increases the efficiency of translation of reovirus s1 mRNA but not s4 mRNA in transfected cells. Virology 176(1):106–113

Schiff LA, Nibert ML, Co MS, Brown EG, Fields BN (1988) Distinct binding sites for zinc and double-stranded RNA in the reovirus outer capsid protein sigma 3. Mol Cell Biol 8(1):273–283

Schmechel S, Chute M, Skinner P, Anderson R, Schiff L (in press) Preferential translation of reovirus mRNA by a sigma3-dependent mechanism. Virology 232:62–73

Sharpe AH, Fields BN (1982) Reovirus inhibition of cellular RNA and protein synthesis: role of the S4 gene. Virology 122(2):381–391

Shatkin AJ, LaFiandra AJ (1972) Transcription by infectious subviral particles of reovirus. J Virol 10(4):698–706

Shepard DA, Ehnstrom JG, Schiff LA (1995) Association of reovirus outer capsid proteins sigma 3 and mu 1 causes a conformational change that renders sigma 3 protease sensitive. J Virol 69(12):8180–8184

Shepard DA, Ehnstrom JG, Skinner PJ, Schiff LA (1996) Mutations in the zinc-binding motif of the reovirus capsid protein sigma3 eliminate its ability to associate with capsid protein mu1. J Virol 70(3):2065–2068

Silverstein SC, Dales S (1968) The penetration of reovirus RNA and initiation of its genetic function in L-strain fibroblasts. J Cell Biol 36:197–230

Silverstein SC, Astell C, Levin DH, Schonberg M, Acs G (1972) The mechanisms of reovirus uncoating and gene activation in vivo. Virology 47:797–806

Skup D, Millward S (1980a) mRNA capping enzymes are masked in reovirus progeny subviral particles. J Virol 34:490–496

Skup D, Millward S (1980b) Reovirus-induced modification of cap-dependent translation in infected L cells. Proc Natl Acad Sci USA 77(1):152–156

LIVERPOOL JOHN MOORES UNIVERSITY
LEARNING SERVICES

Smith RE, Zweerink HJ, Joklik WK (1969) Polypeptide components of virions, top component and cores of reovirus type 3. Virology 39(4):791–810

Spandidos DA, Krystal G, Graham AF (1976) Regulated transcription of the genomes of defective virions and temperature-sensitive mutants of reovirus. J Virol 18:7–19

Sturzenbecker LJ, Nibert M, Furlong D, Fields BN (1987) Intracellular digestion of reovirus particles requires a low pH and is an essential step in the viral infectious cycle. J Virol 61:2351–2361

Tillotson L, Shatkin AJ (1992) Reovirus polypeptide sigma 3 and N-terminal myristoylation of polypeptide mu 1 are required for site-specific cleavage to mu 1C in transfected cells. J Virol 66(4):2180–2186

Tosteson MT, Nibert ML, Fields BN (1993) Ion channels induced in lipid bilayers by subvirion particles of the nonenveloped mammalian reoviruses. Proc Natl Acad Sci USA 90:10549–10552

Walden WE, Godefroy-Colburn T, Thach RE (1981) The role of mRNA competition in regulating translation. I. Demonstration of competition in vivo. J Biol Chem 256(22):11739–11746

Wallis C, Smith KO, Melnick JL (1964) Reovirus activation by heating and inactivation by cooling in MgCl₂ solution. Virology 22:608–619

Ward RL, Shatkin AJ (1972) Association of reovirus mRNA with viral proteins: a possible mechanism for linking the genome segments. Arch Biochem Biophys 152:378–384

Ward R, Banerjee AK, LaFiandra A, Shatkin AJ (1972) Reovirus-specific ribonucleic acid from polysomes of infected L cells. J Virol 9(1):61–69

Watanabe Y, Millward S, Graham AF (1968) Regulation of transcription of the reovirus genome. J Mol Biol 36:107–123

Weiner HL, Ramig RF, Mustoe TA, Fields BN (1978) Identification of the gene coding for the hemagglutinin of reovirus. Virology 86:581–584

Weiner HL, Ault KA, Fields BN (1980) Interaction of reovirus with cell surface receptors. I. Murine and human lymphocytes have a receptor for the hemagglutinin of reovirus type 3. J Immunol 124:2143–2148

Wessner DR, Fields BN (1993) Isolation and genetic characterization of ethanol-resistant reovirus mutants. J Virol 67:2442–2447

White CK, Zweerink HJ (1976) Studies on the structure of reovirus cores: selective removal of polypeptide lambda 2. Virology 70(1):171–180

Yue Z, Shatkin AJ (1996) Regulated, stable expression and nuclear presence of reovirus double-stranded RNA-binding protein s3 in HeLa cells. J Virol 70(6):3497–3501

Zarbl H, Skup D, Millward S (1980) Reovirus progeny subviral particles synthesize uncapped mRNA. J Virol 34(2):497–505

Zweerink HJ (1974) Multiple forms of ss-dsRNA polymerase activity in reovirus-infected cells. Nature (Lond) 247:313–315

Zweerink HJ, Joklik WK (1970) Studies on the intracellular synthesis of reovirus-specified proteins. Virology 41(3):501–518

Zweerink HJ, Ito Y, Matsuhisa T (1972) Synthesis of reovirus double-stranded RNA within virionlike particles. Virology 50(2):349–358

LIVERPOOL JOHN MOORES UNIVERSITY
LEARNING SERVICES

Reovirus σ3 Protein: dsRNA Binding and Inhibition of RNA-Activated Protein Kinase

B.L. Jacobs and J.O. Langland

1 Introduction

The interferon-inducible, double-stranded (ds)RNA-activated protein kinase PKR is clearly an important mediator of interferon action and has also been implicated in the induction of apoptosis (LEE and ESTEBAN 1994; KIBLER et al. 1997) and in the inhibition of cellular protein synthesis in virus-infected cells. Much of the compelling evidence for the role of PKR in virus replication has come from the study of viruses such as reovirus that encode inhibitors of PKR. This review will discuss the genetic and biochemical data suggesting that reovirus σ3 protein, acting as a dsRNA-binding protein, functions as an inhibitor of PKR.

2 The Interferon-Inducible Protein Kinase PKR

The dsRNA-activatable protein kinase PKR has been found in most mammalian cells analyzed. An analogous enzyme, which is immunologically cross-reactive with human PKR, has been identified in plant cells (LANGLAND et al. 1995) and is inducible by virus and viroid infection (CRUM et al. 1988; HIDDINGA et al. 1988; ROTH and HE 1994). In some mammalian cells, PKR expression is induced

Department of Microbiology and the Graduate Program in Molecular and Cellular Biology, Arizona State University, Tempe, AZ 85287-2701, USA severalfold by treatment with either type I or type II interferon. PKR can bind to and be potently activated by dsRNA. Activation occurs concomitantly with inter-molecular (KOSTURA and MATHEWS 1989; THOMIS and SAMUEL 1995) and perhaps intramolecular (BERRY et al. 1985; GALABRU et al. 1989) autophosphorylation, which may be accompanied by dimerization (LANGLAND and JACOBS 1992; PATEL et al. 1995). Once activated, PKR can phosphorylate a number of substrates, including the small (α)-subunit of the protein synthesis initiation factor eIF-2 (FARRELL et al. 1977; LEVIN and LONDON 1978; SAMUEL 1979), the NFκB inhibitor IκB (KUMAR et al. 1994; MARAN et al. 1994; OFFERMANN et al. 1995), and histone proteins (GALABRU and HOVANESSIAN 1985; JACOBS and IMANI 1988). eIF-2α phosphorylation can lead to an inhibition of the initiation of protein synthesis. Phosphorylation of IκB by PKR can lead to its degradation and subsequent activation of NFκB (KUMAR et al. 1994; MARAN et al. 1994; OFFERMANN et al. 1995). The activation of NFκB me-diated by dsRNA may be involved in the induction of interferon-β gene expression and of the other cellular genes whose transcription is influenced by dsRNA. PKR-mediated phosphorylation of histone proteins has only been detected in vitro, and its biological significance is at present unclear.

eIF-2α phosphorylation by PKR is presumed to be involved in the interferon-mediated inhibition of replication of a number of viruses. Constitutive expression of either human (MEURS et al. 1992) or mouse (BAIER et al. 1993) PKR leads to an inhibition of replication of encephalomyocarditis virus (EMCV). For both ade-novirus (KITAJEWSKI et al. 1986) and vaccinia virus (VV) (BEATTIE et al. 1991, 1995a, b; CHANG et al. 1995), deletion of inhibitors of PKR (VAI RNA for ade-novirus and the E3L or K3L genes for VV) leads to increased phosphorylation of eIF-2α and renders these normally interferon-resistant viruses sensitive to the ef-fects of interferon. In the case of adenovirus, deletion of the VAI gene can be complemented by overexpression of a nonphosphorylatable variant of eIF-2α (DAVIES et al. 1989).

Activated PKR may also be able to induce apoptosis in VV-infected cells (KIBLER et al. 1997; LEE and ESTEBAN 1994), although the substrates involved in this response have not been characterized. Inhibition of endogenous PKR, by ex-pression of either dominant-negative mutants of PKR (KOROMILAS et al. 1992; MEURS et al. 1993) or a natural cellular inhibitor of PKR (BARBER et al. 1994), produced a transformed phenotype in cells, as did overexpression of a non-phosphorylatable mutant of eIF-2α (DONZE et al. 1995).

3 σ3 as an Inhibitor of PKR

The identification of the reovirus σ3 protein as an inhibitor of PKR stemmed from two sets of experiments, one biochemical and the other genetic in nature. In 1976, HUISMANS and JOKLIK (1976) identified σ3 as the only reovirus protein in extracts from infected cells that bound specifically to dsRNA. When JACOBS and IMANI

(1988) showed that nonviral dsRNA-binding proteins could act as inhibitors of PKR by sequestering dsRNA, it was a short leap in logic to test σ3 for PKR-inhibitory activity. IMANI and JACOBS (1988) showed that extracts from reovirus T1 (Lang)-infected L cells contained a PKR-inhibitory activity that copurified with σ3. σ3, expressed from a cloned gene, also has PKR-inhibitory activity (DENZLER and JACOBS 1994). All three serotypes of reovirus induce the synthesis of an inhibitor of PKR, although with different kinetics and to different extents (F. IMANI and B.L. JACOBS, unpublished observations). Inhibition of PKR activity seems to be due to binding and sequestering of dsRNA, since PKR inhibition can be overcome by adding excess dsRNA (IMANI and JACOBS 1988), since σ3 acts in a noncatalytic manner (F. IMANI and B.L. JACOBS, unpublished observations), and since PKR inhibition by mutants of σ3 correlates with binding to dsRNA (DENZLER and JACOBS 1994).

Despite encoding an inhibitor of PKR, reoviruses have been reported to be sensitive to the antiviral effects of interferon (WIEBE and JOKLIK 1975). Analysis of several serotypes and strains of reovirus has demonstrated that reovirus T3 (Dearing) may be unique in being sensitive to interferon treatment; T1 (Lang) appears to be quite resistant to interferon treatment, both in single-step growth assays and plaque reduction assays (JACOBS and FERGUSON 1991). Surprisingly, even strains of T3 (Dearing) in some laboratories seem to be relatively resistant to treatment of cells with interferon (L. SCHIFF, personal communication). This, along with the inability to replace genes in reovirus, has made genetic mapping of in-terferon resistance difficult. While it is tempting to assume that the relative resis-tance of reovirus to treatment of cells with interferon is due to the presence of an inhibitor of PKR, there is as yet no direct proof of this hypothesis. In fact, the best evidence for σ3 functioning in infected cells as an inhibitor of PKR comes from heterologous systems, studying the ability of σ3 to replace known inhibitors of PKR. The vaccinia virus (VV) E3L gene encodes a dsRNA-binding protein in-hibitor of PKR (CHANG et al. 1992). Wild-type (wt) VV is interferon resistant and replicates in many cell lines in culture, while VV deleted for E3L has only been shown to replicate in chick embryo fibroblast (CEF) and rabbit RK-13 cells (BEATTIE et al. 1995a, 1996). Replication in RK-13 cells is sensitive to pretreatment of cells with interferon (CHANG et al. 1995). Virus deleted for E3L also induces apoptosis in HeLa cells independent of interferon treatment (LEE and ESTEBAN 1994; KIBLER et al. 1997). In permissive RK-13 cells, virus deleted for E3L induces apoptosis only after interferon treatment (KIBLER et al. 1997). Virus deleted for E3L leads to activation of PKR in infected cells and increased levels of phos-phorylated eIF-2α (BEATTIE et al. 1995a). VV containing σ3 in place of E3L is relatively resistant to treatment of cells with interferon, has an extended host range compared to virus deleted for E3L (BEATTIE et al. 1995a), does not induce apoptosis (KIBLER et al. 1997), and prevents activation of PKR (SEIDLER-WULFF and JACOBS, unpublished observations). In a likewise manner, the adenovirus VAI RNA is necessary for inhibition of PKR during adenovirus infection (KITAJEWSKI et al. 1986). Infection of 293 cells with adenovirus deleted for the VAI RNA gene leads to phosphorylation of eIF-2α and inhibition of protein synthesis. Expression of σ3 in *trans* decreased phosphorylation of eIF-2α and partially restored adenovirus protein synthesis in infected cells (LLOYD and SHATKIN 1992).

The ability of σ3 to prevent the dsRNA-mediated induction of apoptosis in VV-infected cells (KIBLER et al. 1997) suggests that σ3 might be playing a similar role during reovirus infection. Reovirus infection does in fact induce apoptosis in some cells, in a strain-dependent manner (TYLER et al. 1996; RODGERS et al. 1997). However, induction of apoptosis maps primarily to the S1 gene, suggesting that induction of apoptosis might be due to binding of virus to cells. However, the M2 gene, which encodes the μ1 protein, also plays a role in strain differences in induction of apoptosis (TYLER et al. 1996). Since μ1 binding to σ3 can modulate σ3 binding to dsRNA (TILLOTSON and SHATKIN 1992; SHEPARD et al. 1995), it is possible that activators and inhibitors of PKR may play a role in induction of apoptosis in reovirus-infected cells.

Genetic and molecular genetic studies have also indicated that σ3 is crucial in the regulation of protein synthesis in cells. Original studies in Fields' laboratory showed that strains of reovirus differed in their ability to rapidly inhibit host protein synthesis after infection at a high multiplicity of infection (MOI) (SHARPE and FIELDS 1982). T2 (Jones) inhibited host protein synthesis to a much greater extent than T3 (Dearing) or T1 (Lang). The ability to rapidly inhibit host protein synthesis mapped to the σ3-encoding S4 gene (SHARPE and FIELDS 1982). In an attempt to develop a system to study this phenomenon, GIANTINI and SHATKIN (1989) showed that expression of σ3 from a plasmid led to stimulation, rather than inhibition, of expression of a cotransfected reporter gene. This was reminiscent of the translational stimulation of a reporter gene seen upon coexpression of another inhibitor of PKR, the adenovirus VAI RNA (KAUFMAN et al. 1989). VAI RNA has been shown to inhibit activation of PKR by dsRNA synthesized from the transfected plasmid (DAVIES et al. 1989). Again, it was a short mental leap to show that σ3 was indeed acting in transfected cells as an inhibitor of PKR (GIANTINI and SHATKIN 1989) and that σ3 could in fact complement adenovirus deleted of its PKR inhibitor, VAI RNA (LLOYD and SHATKIN 1992). σ3 from each of the three standard laboratory strains of reovirus was able to stimulate translation of an indicator gene, although to different extents. σ3 from T1 (Lang) was the weakest stimulator of translation of the three (SELIGER et al. 1992; MARTIN and McCRAE 1993). Stimulation of translation correlated with accumulation of high levels of σ3, possibly a phenomenon of autostimulation of translation of S4 mRNA by σ3 protein. Stimulation of translation mapped to the C-terminal half of σ3 (MARTIN and McCRAE 1993), the region of σ3 that has also been shown to be necessary for binding to dsRNA.

While there is much evidence suggesting that σ3 may act as a dsRNA-binding protein inhibitor of PKR in reovirus-infected cells, there is really no conclusive evidence as to what RNA it might be binding. Completely uncoated dsRNA genome has not been detected in cells infected with dsRNA viruses (SCHIFF and FIELDS 1990). For the reoviruses, input viral dsRNA remains within the inner capsid throughout the viral life cycle, and progeny genome is only synthesized after assembly of positive-sense single-stranded progeny RNA into subviral particles. It is likely that the machinations that dsRNA viruses go through to prevent exposure of naked dsRNA in cells is a consequence of the profound effects that dsRNA has on the physiology of the cell. Nonetheless, it could be that minute amounts of incorrectly uncoated or packaged genome might be present in infected cells and could activate PKR. Consistent with this suggestion is the fact that a single molecule of dsRNA can induce a quantum response in cells (MARCUS and SEKELLICK 1977). Despite the fact that free dsRNA has not been detected in cells infected with dsRNA viruses, reo- and rotaviruses code for unrelated dsRNA-binding proteins. In the case of group C rotaviruses, the dsRNA-binding protein is nonstructural, suggesting that it functions as a dsRNA-binding protein in the infected cell. Alternatively, for reovirus, secondary structure on mRNA might be involved in activating PKR in infected cells. Such structure has been implicated in the poor translation of the reovirus S1 mRNA (HENRY et al. 1994).

4 Structural Analysis of σ3 Protein

The σ3 protein serves a number of distinct roles in the reovirus life cycle. Differences in inhibition of host RNA and protein synthesis map to the gene encoding σ3 (SHARPE and FIELDS 1982). Mutations in σ3 also accumulate during persistent infection with reovirus (AHMED and FIELDS 1982) and during intertypic reassortment (RONER et al. 1995). These diverse functions in infected cells may be mediated by the diverse biochemical activities of σ3. As described above, a large body of work supports the dsRNA-binding activity associated with σ3 in preventing activation of PKR, thereby stimulating protein synthesis. σ3 protein also associates with the polypeptide μ1 and its cleavage product μ1C to form the outer shell of the mature virion (JAYASURIYA et al. 1988). σ3 is also needed for cleavage of μ1 to μ1C (TILLOTSON and SHATKIN 1992; MABROUK and LEMAY 1994).

The multiple roles of the σ3 protein are likely responsible for the extreme evolutionary conservation of the σ3 amino acid sequence. Sequence comparison of the S4 gene from the three reovirus serotypes suggest that T1 and T3 are more closely related to each other than to T2 (SELIGER et al. 1992). This result is consistent with the suggestion that T2 virus diverged from a progenitor reovirus before T1 and T3 (WIENER and JOKLIK 1988). Most of the nucleotide mismatches occur in the third base position of codons; thus the amino acid sequence is more similar than the gene sequence. For example, T1 and T3, T2 and T3, and T2 and T1 σ3 polypeptides are 97%, 90%, and 90% identical, respectively, while the corresponding S4 gene pairs are 93%, 77%, and 78% identical, respectively. KEDL et al. (1995) have found similar levels of conservation in the σ3 protein from 16 T1 and T3 field isolates, having an average amino acid sequence identity of 96.2%.

The structural and regulatory functions of σ3 are not well understood at a molecular level. Mapping studies have been hampered by the fact that even small mutations at either the N or C terminus destroy dsRNA-binding activity in cells in culture (DENZLER and JACOBS 1994). On a gross level, the amino-terminal region of σ3 is likely involved in μ1/μ1C interaction, whereas dsRNA-binding involves the carboxy terminus. These regions correspond to putative functional domains, an N-terminal zinc finger motif and two C-terminal basic domains (Fig. 1). Mutations in the C-terminal basic domains that inactivate dsRNA binding often have little affect on binding of σ3 to μ1/μ1C (SHEPARD et al. 1996). Alternatively, mutations in the N terminus of σ3, which affect zinc binding and association with μ1/μ1C, have little affect on binding to dsRNA. In fact, some N-terminal mutations increase binding of σ3 to dsRNA in northwestern blot assays (MILLER and SAMUEL 1992; WANG et al. 1996).

The major surface component of reovirus virions consists of σ3–μ1C complexes (JAYASURIYA et al. 1988). Processing of the precursor μ1 to μ1C requires the association of σ3 (LEE et al. 1981; TILLOTSON and SHATKIN 1992). σ3 is the only viral component necessary for processing μ1. In addition to the zinc finger motif, sequences similar to the putative catalytic site of the picornavirus 2A and 3C proteases have been identified in σ3, overlapping with the zinc finger domain (GIANTINI et al. 1984; SCHIFF et al. 1988). However, mutation of these sequences did not affect the ability of σ3 to facilitate processing of μ1 (MABROUK and LEMAY 1994). It is not known whether μ1 cleavage is catalyzed by σ3 directly or whether binding to μ1 exposes the cleavage site to attack by a cellular protease. The picornaviral 2A and 3C proteases are thiol proteases and are subject to inhibition by zinc, presumably binding to the active site cysteine (PELHAM 1978). Therefore, the possibility exists that σ3 may also possess proteolytic activity which may be regulated by binding of zinc to σ3.

Different reovirus serotypes inhibit host protein synthesis to different extents and with different kinetics (SHARPE and FIELDS 1982). The association of σ3 with μ1/μ1C results in the formation of a complex which does not bind dsRNA (TILLOTSON and SHATKIN 1992; SHEPARD et al. 1996). Since the binding sites on σ3 for μ1 and dsRNA do not appear to overlap, it is likely that binding of μ1 by σ3 either interferes with σ3's ability to bind dsRNA in a steric manner or that σ3–μ1 interaction induces a conformational change in σ3 which may render σ3 unable to bind dsRNA. Binding of σ3 to μ1 does induce a conformational change that renders σ3 more accessible to proteases in vitro (SHEPARD et al. 1995). Given the high level of sequence conservation of σ3 between reovirus serotypes, it has been suggested that serotype differences with respect to the inhibition of cellular protein synthesis are likely not related to dsRNA affinities. Alternatively, σ3 binding to

Fig. 1. Domains of σ3. *Arrow,* V8 protease cleavage site (between amino acids 217 and 218); *white box,* 2A/3C protease homology (amino acids 45–53); *diagonal shading,* zinc finger motif (amino acids 51–71); *black box,* basic domain (amino acids 234–240 and 291–297); *vertical shading,* PP2A subunit A homology (amino acids 240–267)

dsRNA may be modulated by the amount of free σ3 which is capable of binding dsRNA. Consequently, the relative expression levels of σ3 and μ1 at critical times after infection may be an important regulatory mechanism.

The motif within σ3 responsible for the dsRNA-binding activity has not been fully identified. This is partly due to differing results that have been obtained when measuring binding by a northwestern assay as opposed to measuring binding of soluble protein. Northwestern blot analysis of proteolytic fragments localized the site of dsRNA-binding activity to the C-terminal 16-kDa fragment (Fig. 1) of the full-length 41-kDa protein (SCHIFF et al. 1988; MILLER and SAMUEL 1992). Deletion of the N terminus actually stimulated dsRNA-binding activity. A similar analysis has not been possible with soluble σ3, since even small deletions from the N or C terminus destroy binding to dsRNA (DENZLER and JACOBS 1994). The region identified by northwestern analysis is distinct from the zinc finger motif and does not possess appreciable sequence homology to other dsRNA-binding proteins. This region of σ3 contains two basic domains (amino acids 233–234 and 286–305) that are predicted to fold as α-helices (YUE and SHATKIN 1996). A similar structure (BYCROFT et al. 1995; KHARRAT et al. 1995) has been identified for a dsRNA-binding domain conserved among several dsRNA-binding proteins unrelated to σ3 (MCCORMACK et al. 1992; CHANG and JACOBS 1993; ST. JOHNSTON et al. 1993). The magnetic resonance imaging (MRI) structure of this domain appears to consist of an α-β-β-β-α topology in which the α-helices are present on the one face of a three-stranded antiparallel β-sheet (BYCROFT et al. 1995; KHARRAT et al. 1995). This arrangement likely facilitates the electrostatic interaction of key basic residues with the negatively charged surface of dsRNA. It is therefore possible that, while σ3 shares no sequence homology with proteins that contain a conserved dsRNA-binding motif, it might bind dsRNA through a structurally similar element.

Analysis of point mutations of basic residues in the C-terminal domains of σ3 confirm the importance of these basic domains in dsRNA binding (DENZLER and JACOBS 1994; MABROUK et al. 1995; WANG et al. 1996). In soluble assays, substitutions at Arg-236, Arg-239, Lys-291, and Lys-293 eliminated the ability of σ3 to bind dsRNA. For Lys-291 and Lys-293, Arg substitutions restored dsRNA-binding ability, suggesting that the positive charge at this position is important. Nonconserved mutations at Lys-234, Lys-240, Arg-296, and Lys-297 did not affect the protein's ability to bind dsRNA (DENZLER and JACOBS 1994). In addition, with soluble assays, a deletion mutation which could disrupt the formation of an α-helix prevented dsRNA binding (DENZLER and JACOBS 1994). The importance of the two basic domains has been confirmed with the northwestern blot assay, although the more C-terminal basic domain seems to be necessary for binding only in the presence of the inhibitory N-terminal half of σ3 (WANG et al. 1996). In contrast to results with soluble assays, with the northwestern blot assay, point mutations that were predicted to interrupt α-helix formation had little effect on dsRNA binding (WANG et al. 1996). In support of these mutational observations, no amino acid differences exist among the three laboratory strains within the first basic motif (amino acids 233–243) suggested to be involved in dsRNA binding. Within the second motif (amino acids 286–05), only conserved amino acid changes are observed between serotypes. Specifically, T1 and T3 contain a Lys at residue 293 and an Arg at residue 296, whereas T2 contains an Arg and Lys, respectively (SELIGER et al. 1992). Near the basic domains, at amino acid 344, a nonconserved amino acid change between serotypes is observed where a Ser in T2 and T3 is substituted with a Pro in T1 (SELIGER et al. 1992). This change could have important effects on polypeptide folding in this region. Such altered folding could account for differences observed between the ability of σ3 from T2 and T3 to stimulate chloramphenicol acetyltransferase (CAT) expression in cotransfected COS cells as compared to T1 (SELIGER et al. 1992; MARTIN and McCRAE 1993).

Viral replication and the regulation of protein synthesis in VV-infected cells appears to be dependent on the expression of dsRNA-binding proteins from the E3L gene (JACOBS and LANGLAND 1996). VV deleted of the E3L gene is unable to form plaques on Vero cells, and replication is inhibited in interferon-treated RK-13 and L cells (BEATTIE et al. 1995a; CHANG et al. 1995). Expression of the reovirus S4 gene is able to rescue the replication of VV deleted for E3L, and the ability of σ3 to rescue replication correlates with the ability of σ3 mutants to bind dsRNA (DENZLER and JACOBS 1994; BEATTIE et al. 1995a). These data further support the role of the basic domains in dsRNA binding.

One additional region of sequence homology has been noted for σ3. A region between the two basic domains (Fig. 1), extending from amino acid 240 to 267, shows a high degree of sequence similarity to the 65-kDa regulatory A subunit of the protein phosphatase PP2A (71% similarity and 39% identity) (MILLER and SAMUEL 1992). PP2A has been identified as the eIF-2α protein phosphatase, and association of the A subunit increases the rate of eIF-2α dephosphorylation. Consequently, it is possible that σ3 binds to and modulates the activity of PP2A, altering protein synthesis in the reovirus-infected cell.

5 Concluding Remarks

The reoviral inhibition of PKR is unique among the viral inhibitors of PKR in that the interaction of σ3 with other viral, and perhaps even cellular, macromolecules can control its activity. While the myriads of interactions that may control σ3 function are fascinating, they have made this system difficult to unravel. However, even the small amount of insight that we have obtained into the role of σ3 in translation in infected cells gives us hope that finally unraveling this system will likely tell us a great deal about the interactions between reovirus and its host.

Acknowledgements. This review is dedicated to the memory of Bernie Fields, whose encouragement and support of work concerning reovirus, PKR, and interferon has been instrumental in moving the field forward. This work was supported by Public Health Service grant CA-48654 from the National Cancer Institute, contract CNTR 9610 from the Arizona Disease Control Research Commission, and grant VM-151 from the American Cancer Society.

References

Ahmed R, Fields BN (1982) Role of the S4 gene in the establishment of persistent reovirus infection in L cells. Cell 28:605–612

Baier L, Shors T, Shors ST, Jacobs BL (1993) The mouse phosphotyrosine immunoreactive kinase, TIK, is indistinguishable from the double-stranded RNA-dependent, interferon-inducible protein kinase, PKR. Nucleic Acids Res 21:4830–4835

Barber GN, Thompson S, Lee TG, Strom T, Jagus R, Darveau A, Katze MG (1994) The 58-kilodalton inhibitor of the interferon-induced double-stranded RNA-activated protein kinase is a tetratricopeptide repeat protein with oncogenic properties. Proc Natl Acad Sci USA 91:4278–4282

Beattie E, Tartaglia J, Paoletti E (1991) Vaccinia virus-encoded eIF-2a homologue abrogates the antiviral effect of interferon. Virology 183:419–422

Beattie E, Denzler K, Tartaglia J, Perkus ME, Paoletti E, Jacobs BL (1995a) Reversal of the interferon-sensitive phenotype of a vaccinia virus lacking E3L by expression of the reovirus S4 gene. J Virol 69:499–505

Beattie E, Paoletti E, Tartaglia J (1995b) Distinct patterns of IFN sensitivity observed in cells infected with vaccinia K3L- and E3L-mutant viruses. Virology 210:254–263

Beattie E, Kaufman E, Martinez H, Perkus M, Jacobs BL, Paoletti E, Tartaglia J (1996) Host range restriction of vaccinia virus E3L-specific deletion mutants. Virus Genes 12:89–94

Berry MJ, Knutson GS, Laskey SR, Munemitsu SM and Samuel CE (1985). Mechanism of action of interferon. Purification and substrate specificities of the double-stranded RNA-dependent protein kinase from untreated and interferon-treated mouse fibroblasts. J Biol Chem 260:11240–11247

Bycroft M, Proctor M, Freund SM, St Johnston D (1995) Assignment of the backbone 1H,15N,13C NMR resonances and secondary structure of a double-stranded RNA binding domain from the Drosophila protein staufen. FEBS Lett 362:333–336

Chang H-W, Jacobs BL (1993) Identification of a conserved motif that is necessary for binding of the vaccinia virus E3L gene products to double-stranded RNA. Virology 194:537–547

Chang H-W, Watson J, Jacobs BL (1992) The vaccinia virus E3L gene encodes a double-stranded RNA-binding protein with inhibitory activity for the interferon-induced protein kinase. Proc Natl Acad Sci USA 89:4825–4829

Chang HW, Uribe LH, Jacobs BL (1995) Rescue of vaccinia virus lacking the E3L gene by mutants of E3L. J Virol 69:6605–6608

Choi SY, Scherer BJ, Schnier J, Davies MV, Kaufman RJ, Hershey JW (1992) Stimulation of protein synthesis in COS cells transfected with variants of the alpha-subunit of initiation factor eIF-2. J Biol Chem 267:286–293

Crum CJ, Hiddinga HJ, Roth DA (1988) Tobacco mosaic virus infection stimulates the phosphorylation of a plant protein associated with double-stranded RNA-dependent protein kinase activity. J Biol Chem 263:13440–13443

Davies MV, Furtado M, Hershey JW, Thimmappaya B, Kaufman RJ (1989) Complementation of adenovirus virus-associated RNA I gene deletion by expression of a mutant eukaryotic translation initiation factor. Proc Natl Acad Sci USA 86:9163–9167

Degen HJ, Blum D, Grun J, Jungwirth C (1992) Expression of authentic vaccinia virus-specific and inserted viral and cellular genes under control of an early vaccinia virus promoter is regulated posttranscriptionally in interferon-treated chick embryo fibroblasts. Virology 188:114–121

Denzler K, Jacobs BL (1994) Site-directed mutagenic analysis of reovirus σ3 binding to dsRNA. Virology 204:190–199

Donze O, Jagus R, Koromilas AE, Hershey JW, Sonenberg N (1995) Abrogation of translation initiation factor eIF-2 phosphorylation causes malignant transformation of NIH 3T3 cells. EMBO J 14:3828–3834

Farrell PJ, Balkow K, Hunt T, Jackson RJ, Trachsel H (1977) Phosphorylation of initiation factor eIF-2 and control of reticulocyte protein synthesis. Cell 11:187–200

Galabru J, Hovanessian AG (1985) Two interferon-induced proteins are involved in the protein kinase complex dependent on double-stranded RNA. Cell 43:685–694

Galabru J, Katze MG, Robert N and Hovanessian AG (1989). The binding of double-stranded RNA and adenovirus VAI RNA to the interferon-induced protein kinase. Eur. J. Biochem. 178:581–589

Giantini M, Shatkin AJ (1989) Stimulation of chloramphenicol acetyltransferase mRNA translation by reovirus capsid polypeptide sigma 3 in cotransfected COS cells. J Virol 63:2415–2421

Giantini M, Seliger LS, Furuichi Y, Shatkin AJ (1984) Reovirus type 3 genome segment S4: nucleotide sequence of the gene encoding a major virion surface protein. J Virol 52:984–987

Grun J, Redmann Muller I, Blum D, Degen HJ, Doenecke D, Zentgraf HW, Jungwirth C (1991) Regulation of histone H5 and H1 zero gene expression under the control of vaccinia virus-specific sequences in interferon-treated chick embryo fibroblasts. Virology 180:535–542

Henry GL, McCormack SJ, Thomis DC, Samuel CE (1994) Mechanism of interferon action. Translational control and the RNA-dependent protein kinase (PKR): antagonists of PKR enhance the translational activity of mRNAs that include a 161 nucleotide region from reovirus S1 mRNA. J Biol Regul Homeost Agents 8:15–24

Hiddinga HJ, Crum CJ, Hu J, Roth DA (1988) Viroid-induced phosphorylation of a host protein related to the dsRNA-dependent protein kinase. Science 241:451–453

Huismans H, Joklik WK (1976) Reovirus-coded polypeptides in infected cells: isolation of two native monomeric polypeptides with affinity for single-stranded and double-stranded RNA, respectively. Virology 70:411–424

Imani F, Jacobs BL (1988) Inhibitory activity for the interferon-induced protein kinase is associated with the reovirus serotype 1 sigma 3 protein. Proc Natl Acad Sci USA 85:7887–7891

Jacobs BL, Ferguson RE (1991) Reovirus serotypes 1 and 3 differ in their sensitivity to interferon. J Virol 65:5102–5104

Jacobs BL, Imani F (1988) Histone proteins inhibit activation of the interferon-induced protein kinase by binding to double-stranded RNA. J Interferon Res 8:821–830

Jacobs BL, Langland JO (1996) When two strands are better than one: the mediators and modulators of the cellular responses to double-stranded RNA. Virology 219:339–349

Jagus R, Gray MM (1994) Proteins that interact with PKR. Biochimie 76:779–791

Jayasuriya AK, Nibert ML, Fields BN (1988) Complete nucleotide sequence of the M2 gene segment of reovirus type 3 Dearing and analysis of its protein product mu 1. Virology 163:591–602

Kaufman RJ, Davies MV, Pathak VK, Hershey JW (1989) The phosphorylation state of eucaryotic initiation factor 2 alters translational efficiency of specific mRNAs. Mol Cell Biol 9:946–958

Kedl R, Schmechel S, Schiff L (1995) Comparative sequence analysis of the reovirus S4 genes from 13 serotype 1 and serotype 3 field isolates. J Virol 69:552–559

Kharrat A, Macias MJ, Gibson TJ, Nilges M, Pastore A (1995) Structure of the dsRNA binding domain of E. coli RNase III. EMBO J 14:3572–3584

Kibler KV, Shors T, Perkins KB, Zeman CC, Banaszak MP, Biesterfeldt J, Langland JO, Jacobs BL (1997) Double-stranded RNA is a trigger for apoptosis in vaccinia virus-infected cells. J Virol 71:1992–2003

Kitajewski J, Schneider RJ, Safer B, Munemitsu SM, Samuel CE, Thimmappaya B, Shenk T (1986) Adenovirus VAI RNA antagonizes the antiviral action of interferon by preventing activation of the interferon-induced eIF-2 kinase. Cell 45:195–200

Koromilas AE, Roy S, Barber GN, Katze, MG, Sonenberg N (1992) Malignant transformation by a mutant of the IFN-inducible dsRNA-dependent protein kinase. Science 257:1685–1689

Kostura M, Mathews MB (1989) Purification and activation of the double-stranded RNA-dependent eIF-2 kinase DAI. Mol Cell Biol 9:1576–1586

Kumar A, Haque J, Lacoste J, Hiscott J, Williams BR (1994) Double-stranded RNA-dependent protein kinase activates transcription factor NF-kappa B by phosphorylating I kappa B. Proc Natl Acad Sci USA 91:6288–6292

Langland JO, Jacobs BL (1992) Cytosolic double-stranded RNA-dependent protein kinase is likely a dimer of partially phosphorylated Mr = 66,000 subunits. J Biol Chem 267:10729–10736

Langland JO, Song J, Jacobs BL, Roth DA (1995) A plant-encoded analogue of PKR, the mammalian double-stranded RNA-dependent protein kinase. Plant Physiol 108:1259–1267

Lee PW, Hayes EC, Joklik WK (1981) Characterization of anti-reovirus immunoglobulins secreted by cloned hybridoma cell lines. Virology 108:134–146

Lee SB, Esteban M (1994) The interferon-induced double-stranded RNA-activated protein kinase induces apoptosis. Virology 199:491–496

Levin D, London IM (1978) Regulation of protein synthesis: activation by double-stranded RNA of a protein kinase that phosphorylates eukaryotic initiation factor 2. Proc Natl Acad Sci USA 75:1121–1125

Lloyd RM, Shatkin AJ (1992) Translational stimulation by reovirus polypeptide sigma 3: substitution for VAI RNA and inhibition of phosphorylation of the alpha subunit of eukaryotic initiation factor 2. J Virol 66:6878–6884

Mabrouk T, Lemay G (1994) The sequence similarity of reovirus sigma 3 protein to picornaviral proteases is unrelated to its role in mu 1 viral protein cleavage. Virology 202:615–620

Mabrouk T, Danis C, Lemay G (1995) Two basic motifs of reovirus sigma 3 protein are involved in double-stranded RNA binding. Biochem Cell Biol 73:137–145

Maran A, Maitra RK, Kumar A, Dong B, Xiao W, Li G, Williams BR, Torrence PF, Silverman RH (1994) Blockage of NF-kappa B signaling by selective ablation of an mRNA target by 2-5A antisense chimeras. Science 265:789–792

Marcus PI, Sekellick MJ (1977) Defective interfering particles with covalently linked [+/−]RNA induce interferon. Nature 266:815–819

Martin PE, McCrae MA (1993) Analysis of the stimulation of reporter gene expression by the sigma 3 protein of reovirus in co-transfected cells. J Gen Virol 74:1055–1062

McCormack SJ, Thomis DC, Samuel CE (1992) Mechanism of interferon action: identification of a RNA binding domain within the N-terminal region of the human RNA-dependent P1/eIF-2 alpha protein kinase. Virology 188:47–56

Meurs EF, Watanabe Y, Kadereit S, Barber GN, Katze MG, Chong K, Williams BRG, Hovanessian AG (1992) Constitutive expression of human double-stranded RNA-activated p68 kinase in murine cells mediates phosphorylation of eukaryotic initiation factor 2 and partial resistance to encephalomyocarditis growth. J Virol 66:5805–5814

Meurs EF, Galabru J, Barber GN, Katze MG, Hovanessian AG (1993) Tumor suppressor function of the interferon-induced double-stranded RNA-activated protein kinase. Proc Natl Acad Sci USA 90:232–236

Miller JE, Samuel CE (1992) Proteolytic cleavage of the reovirus sigma 3 protein results in enhanced double-stranded RNA-binding activity: identification of a repeated basic amino acid motif within the C-terminal binding region. J Virol 66:5347–5356

Offermann MK, Zimring J, Mellits KH, Hagan MK, Shaw R, Medford RM, Mathews MB, Goodbourn S, Jagus R (1995) Activation of the double-stranded-RNA-activated protein kinase and induction of vascular adhesion molecule-1 by poly(I) · poly(C) in endothelial cells. Eur J Biochem 232:28–36

Patel RC, Stanton P, McMillan NM, Williams BR, Sen GC (1995) The interferon-inducible double-stranded RNA-activated protein kinase self-associates in vitro and in vivo. Proc Natl Acad Sci USA 92:8283–8287

Pelham HR (1978) Translation of encephalomyocarditis virus RNA in vitro yields an active proteolytic processing enzyme. Eur J Biochem 85:457–462

Ramaiah KV, Davies MV, Chen JJ, Kaufman RJ (1994) Expression of mutant eukaryotic initiation factor 2 alpha subunit (eIF-2 alpha) reduces inhibition of guanine nucleotide exchange activity of eIF–2B mediated by eIF-2 alpha phosphorylation. Mol Cell Biol 14:4546–4553

Rodgers SE, Barton ES, Oberhaus SM, Pike B, Gibson CA, Tyler KL, Dermody TS (1997) Reovirus-induced apoptosis of MDCK cells is not linked to viral yield and is blocked by Bcl-2. J Virol 71:2540–2546

Roner MR, Lin PN, Nepluev I, Kong LJ, Joklik WK (1995) Identification of signals required for the insertion of heterologous genome segments into the reovirus genome. Proc Natl Acad Sci USA 92:12362–12366

Roth DA, He X (1994) Viral-dependent phosphorylation of a dsRNA-dependent kinase. Prog Mol Subcell Biol 14:28–47

Samuel CE (1979) Mechanism of interferon action: phosphorylation of protein synthesis initiation factor eIF-2 in interferon-treated human cells by a ribosome-associated kinase possessing site specificity similar to hemin-regulated rabbit reticulocyte kinase. Proc Natl Acad Sci USA 76:600–604

Schiff LA, Fields BN (1990) Reoviruses and their replication. In: Fields BN, Knipe DM et al. (eds) Fundamental virology, 2nd edn. Raven, New York, pp 583–618

Schiff LA, Nibert ML, Co MS, Brown EG, Fields BN (1988) Distinct binding sites for zinc and double-stranded RNA in the reovirus outer capsid protein sigma 3. Mol Cell Biol 8:273–283

Seliger LS, Giantini M, Shatkin AJ (1992) Translational effects and sequence comparisons of the three serotypes of the reovirus S4 gene. Virology 187:202–210

Sharpe AH, Fields BN (1982) Reovirus inhibition of cellular RNA and protein synthesis: role of the S4 gene. Virology 122:81–391

Shepard DA, Ehnstrom JG, Schiff LA (1995) Association of reovirus outer capsid proteins sigma 3 and mu 1 causes a conformational change that renders sigma 3 protease sensitive. J Virol 69:8180–8184

Shepard DA, Ehnstrom JG, Skinner PJ, Schiff LA (1996) Mutations in the zinc-binding motif of the reovirus capsid protein delta 3 eliminate its ability to associate with capsid protein mu 1. J Virol 70:2065–2068

St. Johnston D, Brown NH, Gall JG, Jantsch M (1993) A conserved double-stranded RNA binding domain. Proc Natl Acad Sci USA 89:10979–10983

Thomis DC, Samuel CE (1995) Mechanism of interferon action: characterization of the intermolecular autophosphorylation of PKR, the interferon-inducible, RNA-dependent protein kinase. J Virol 69:5195–5198

Tillotson L, Shatkin AJ (1992) Reovirus polypeptide sigma 3 and N-terminal myristoylation of polypeptide mu 1 are required for site-specific cleavage to mu 1C in transfected cells. J Virol 66:2180–2186

Toczyski DP, Steitz JA (1991) EAP, a highly conserved cellular protein associated with Epstein-Barr virus small RNAs (EBERs). EMBO J 10:459–466

Tyler KL, Squier MK, Brown AL, Pike B, Willis D, Oberhaus SM, Dermody TS, Cohen JJ (1996) Linkage between reovirus-induced apoptosis and inhibition of cellular DNA synthesis: role of the S1 and M2 genes. J Virol 70:7984–7991

Wang Q, Bergeron J, Mabrouk T, Lemay G (1996) Site-directed mutagenesis of the double-stranded RNA binding domain of bacterially-expressed sigma 3 reovirus protein. Virus Res 41:141–151

Wiebe ME, Joklik TW (1975) The mechanism of inhibition of reovirus replication by interferon. Virology 66:229–240

Wiener JR, Joklik WK (1988) Evolution of reovirus genes: a comparison of serotype 1, 2, and 3 M2 genome segments, which encode the major structural capsid protein mu 1C. Virology 163:603–613

Yue Z, Shatkin AJ (1996) Regulated, stable expression and nuclear presence of retrovirus double-stranded RNA-binding protein sigma3 in HeLa cells. J Virol 70:3497–3501

Reovirus M1 Gene Expression

E.G. Brown

1 Introduction

The μ2 protein, along with the polymerase protein λ3, is present in catalytic amounts in virions (approximately 12 copies) and may be positioned at the inner face of the vertices of the inner capsid (NIBERT et al. 1996). Since μ2 protein, along with λ3, is the least abundant viral protein, it was missed in the original studies that examined reovirus protein structure (LOH and SHATKIN 1968; SMITH et al. 1969). These studies employed the then state-of-the-art technology, sodium dodecyl sulfate-polyacrylamide gel electrophoresis (SDS-PAGE) in tube gels that was of lower resolution than the slab gels employed later. Since μ2 protein has an electrophoretic mobility similar to the most abundant reovirus protein μ1c, it was obscured in theses studies. μ2 protein was discovered when reovirus mRNA – obtained from melted viral dsRNA purified by SDS-PAGE – was translated in vitro (BOTH et al. 1975; MCCRAE and JOKLIK 1978).

Department of Microbiology and Immunology, Faculty of Medicine, University of Ottawa, 451 Smyth Road, Ottawa, Ontario, Canada K1H 8M5

In this review, I will address the location of replication and encapsidation signals in the M1 genome segment as well as the expression of recombinant μ2 protein. Analysis of this protein has been lacking, presumably because of its low abundance in virus particles, poor reactivity with anti-reovirus antibody, and the difficulty of detecting μ2 protein by SDS-PAGE. These difficulties have been overcome by expressing μ2 protein via recombinant DNA technology to produce sufficient quantities of protein to produce antisera and to study μ2 expression in eukaryotic cells via cloned genes.

2 Biological Properties of the M1 Gene

The reovirus M1 gene is 2304 nucleotides long and encodes a single open reading frame that is translated into the μ2 protein of 736 amino acids. The specific functions of μ2 protein are unknown; however, recent evidence shows a role in transcription and replication, since the difference in temperature optimum and extent of transcription between T1 Lang and T3 Dearing is controlled by the M1 gene and the M1 plus L1 genes, respectively (YIN et al. 1996), and the M1 gene temperature-sensitive (*ts*) mutant *tsH11.2* is defective in dsRNA synthesis (COOMBS 1996; for a review, see the chapter by K.M. Coombs, this volume). The M1 gene controls the difference in cytopathic effect (CPE) measured as loss of trypan blue exclusion and plaque size for T3 Dearing and T2 Jones, where T3 induces greater plaque size and CPE (MOODY and JOKLIK 1989). This is consistent with μ2 providing a control function, but it has also been hypothesized to indicate a toxic effect. Interestingly, along with L1 and L3, the M1 genome segment determines the increased yield of T1 Lang in cardiac cells, but only L1 and L3 determine the increased growth in L929 cells (MATOBA et al. 1991). The M1 gene has also been shown to be of central importance in virulence and hepatotropism in severe combined immunodeficient (SCID) mice (HALLER et al. 1995). The μ2 protein alone determines the ability of T1 Lang to grow in bovine aortic endothelial cells, where T3 Dearing does not grow (MATOBA et al. 1993). Similarly, μ2 protein alone controls the endocarditic phenotype of an endocarditic variant, B8 (SHERRY and FIELDS 1989). Genetic analysis of different viruses that also induced myocarditis implicated μ2 along with other genes encoding core proteins (SHERRY and BLUM 1994). The genetic basis for reovirus myocarditis is reviewed in greater depth in this series by B. SHERRY (see her chapter in vol. 2). It is clear from these biological and functional assays that μ2 protein plays critical roles in viral growth, tropism, and disease which may be due at least in part to its function in the control of RNA synthesis both at the level of transcription and replication.

3 Signals for Replication and Encapsidation of the M1 Genome Segment

In reovirus, the genetic signals that control genome replication and encapsidation are unknown. For each of the genome segments of reovirus to be replicated and assembled, they must contain both transcriptase and replicase promoters as well as signals for encapsidation. Encapsidation involves the gathering of genome segments such that one copy of each segment is incorporated into each nascent subviral particle. Thus sorting and exclusion mechanisms are required to control the process of genome assembly.

Reovirus replication begins with transcription of plus-strand RNA from each of the ten minus strands of parental RNA within the uncoated parental subviral particles (reviewed by JOKLIK 1974; ZARBL and MILWARD 1983; NIBERT et al. 1996). The ten full-length capped plus-strand copies are assembled together with viral proteins to form the nascent viral core, where minus-strand synthesis occurs upon the plus-strand template (ACS et al. 1971; SUKUMA and WATANABE 1972; ZWEERINK et al. 1972). Although it is known that each reovirion contains one of each of the ten genome segments (MILLWARD and GRAHAM 1970; SPENDLOVE et al. 1970; KAVENOFF et al. 1975), the signals that control the replication of each genome segment and the mechanism that ensures that ten unique RNA segments are assembled together in each nascent particle remains unknown. The terminal regions of all ten genome segments share GCUA at their 5′ end and UCAUC at their 3′ end (ANTCZAK et al. 1982), and adjacent to these common end-specific structures is a segment-specific stretch of nucleotides that is complementary to a stretch at the opposite end of the segment (GAILLARD et al. 1982). Although these short terminal regions are expected to be important for specific functions required in replication and assembly, they are probably not sufficient in themselves to signal all the events in replication and assembly.

Studies with wound tumor virus (WTV), the type member of the genus *Phytoreovirus* of the *Reoviridae* family, suggested that each genome segment must contain at least two operational sequence domains for genome packaging: one that specifies that it is a viral and not a cellular RNA – perhaps the conserved terminal sequences – and a second that specifies that it is a particular RNA genome segment – perhaps the segment-specific terminal inverted repeat sequences (ANZOLA et al. 1987). It was further proposed that the signals required for replication of the WTV genome reside within the terminal domains of the individual segments due to their location and their conservation in deletion mutants (XU et al. 1989; DALL et al. 1990).

Serial passage of viruses at high multiplicity of infection can generate deletion mutants (reviewed by PERRAULT 1981). All genome remnants present in deletion mutants that are derived from linear genomes and have been characterized conserve one or both termini of the virus (reviewed by SCHLESINGER 1988). Reovirus T3 Dearing normally generates deletion mutations in genome segments L1 and L3 (AHMED et al. 1980a, b; BROWN et al. 1983). The temperature-sensitive (*ts*) mutant *ts*C(*447*) (AHMED and GRAHAM 1977) and specific T1 × T3 reassortants (BROWN et al. 1983) can produce deletions in the M1 genome segment and, in the latter instance, in the L2 genome segment as well. T1 (Lang) and T3 (Dearing) reassortants (T1 × T3) that possess the T3 L2 segment and T1 M3 segment generate M1 segment deletions on serial passage (Brown et al. 1983). Since genome segments with deletions can still replicate and be assembled into progeny virus, M1 fragments must maintain the signals essential for their replication and assembly and, furthermore, the smallest fragments will basically consist of the minimum genetic signals for replication and assembly.

M1 deletion fragments identified in high-passage stocks of reovirus reassortants by northern blot analysis using probes specific for the 3′ and 5′ termini indicate that all M1 fragments contain both termini. It is thus possible to amplify M1 deletion mutants by polymerase chain reaction (PCR) of cDNA using end-specific primers. Cloning and sequencing of 13 of the smallest fragments showed that the minimal length of the M1 deletion fragments was 344 nucleotides. All 13 M1 deletion fragments conserved 132–135 nucleotides from the 5′ end of the M1 genome segment and 183–185 nucleotides from the 3′ end. Therefore, these two terminal regions contain the minimum sequences essential for replication and encapsidation of this segment. It is not known what part of this sequence harbors specific signals for the recognition events in replication and encapsidation. In addition to specific sequence signals, the deleted fragment might also have a limit in length, and 344 nucleotides could be the minimum length essential for replication and/or encapsidation. Some of the sequence may function as spacers required to separate binding sites that form specific secondary or tertiary structures or multicomponent complexes. It is possible that protein–RNA or RNA–RNA intermediates are formed in replication and that spatial and steric parameters must be satisfied. It may be a general feature of all viruses with segmented RNA genomes that both terminal sequences play critical roles in genome transcription, replication, and packaging (Nuss and Summers 1984; Zou and Brown 1992b).

Deletion mutants of other dsRNA viruses have all conserved their terminal regions. The fragments of two deletion mutants of WTV possessed consensus sequences of 319 nucleotides from the 5′ end and 205 nucleotides from the 3′ end (Anzola et al. 1987). Sequence analysis of a deletion mutant of yeast ScV "killer" dsRNA segment M, termed S-dsRNA (suppressive), showed that it contains 232 nucleotides from the 5′ end and 550 nucleotides from the 3′ end (Theile et al. 1984). The smallest dsRNA fragment described to date is a 315-nucleotide fragment of the polyhedrin gene of cytoplasmic polyhedrosis virus (CPV), a reovirus which infects insects, consisting of 121 nucleotides from the 5′ end and 191 nucleotides from the 3′ end (reported by Nuss 1988). The similarity in size of the regions seen to be conserved in CPV and reovirus suggest that this may be the location of replication and encapsidation signals in all members of the Reoviridae family. Attempts to introduce the neomycin resistance gene into reovirus by flanking it with the defined minimal 3′ and 5′ regions of the M1 gene that are sufficient for replication and encapsidation have been unsuccessful despite extensive attempts employing in vitro and in vivo strategies of synthesis and introduction of synthetic ssRNA and dsRNA analogues into cells infected with helper virus (S. Zou and E.G. Brown, unpublished results).

4 Conservation of M1 Gene Between T1 Lang and T3 Dearing

The T1 and T3 M1 genes consist of 2304 nucleotides and contain the same initiation and termination codons (WIENER et al. 1989, ZOU and BROWN 1992a). Comparison of T1 and T3 M1 sequences showed high homology in nucleotide sequences (2253 out of 2304, 97.79%) and in predicted amino acid sequences (726 out of 736, 98.64%). There are only 51 nucleotide substitutions, of which 39 (76.47%) occur in the third base codon position. Of the 51 substitutions, 76% were in the third base codon position, with 16% and 8% in the first and second base codon positions, respectively. Thus, out of the 2.2% total nucleotide variability observed, 5.3% of the third base positions were substituted relative to 1.1% and 0.5% of the first and second positions, respectively. The low level of third base substitution indicated that the T1 and T3 M1 segments were closely related, and the relatively less variable first and second base codon positions indicated that the $\mu 2$ protein is structurally conserved and thus possesses a high functional density. This is not surprising, since $\mu 2$ is an internal core component that must interact with the other core proteins $\sigma 2$, $\lambda 1$, $\lambda 2$, $\lambda 3$, and possibly the dsRNA genome.

An estimate for evaluating divergence for T1 and T3 genome segments shows that the M1 and L2 segments have the lowest level of third base codon position divergences, 6% and 7%, respectively, suggesting that they have been evolving independently for the shortest time, whereas L1, L2, and S4 have been evolving for an intermediate time, and the remaining genome segments (M2, S1, S2, and S3) for the longest time (ZOU and BROWN 1992a). This is consistent with reassortment-mediated exchange of genetic material that has occurred at different times and that involved groups of genome segments.

There are 51 nucleotide substitutions in the T1 M1 segment relative to T3; one in the noncoding region and 40 others in the coding region did not alter the sense of codons in the M1 segment. Only ten amino acid substitutions were predicted for T1 relative to T3. Of the ten predicted changes, five were conservative. Five nonconservative substitutions are predicted (at amino acid positions 150, 302, 347, 458, and 726). One or other of these is most likely to be responsible for the differences in phenotype of T1 and T3 that have been associated with the M1 genome segment, differences in transcription optima, CPE, plaque size, or the capacity to replicate in specific cell types as well as cause organ-specific disease.

5 $\mu 2$ Protein Expression In Vitro and In Vivo

5.1 Low-Level Expression In Vivo

In studies of reovirus mRNA translation efficiency determined as the amount of protein synthesis per unit of mRNA in infected, actinomycin D-treated BHK cells,

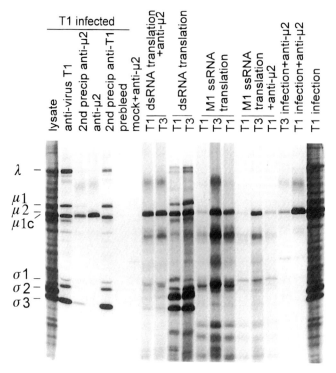

Fig. 1. Radioimmunoprecipitation (RIPA) of μ2 proteins from infected cells, in vitro translation products of M1 cDNA transcripts, and melted reovirion dsRNA. *Left*, differential precipitations of T1-infected L929 lysate first by anti-T1 virus antiserum and then by anti-T1 μ2 and vice versa. *Prebleed*, T1-infected cells precipitated by prebleed from Trp-E-μ2 immunized rabbit; *mock*, L929 cells precipitated with anti Trp-E-T1 μ2 protein. The other lanes show unprecipitated and precipitated in vitro translation products of in vitro T7 transcripts of cloned M1 cDNA or melted viral dsRNA. Two translation reactions of T1M1 ssRNA were included in the gel; the lane on the left is of RNA that was heated to 70 °C for 5 min before translation. (From ZOU and BROWN 1996a)

μ2 expression was shown to be the least efficiently translated viral protein, yielding 100-fold less protein than that of the most efficiently translated S4 gene (GAILLARD and JOKLIK 1985). Low translatability of M1 mRNA is seen most dramatically in vivo, since in vitro translation results in translation efficiencies that are similar to other reoviral proteins (Fig. 1). Since it is known that sequences around the initiation codon, including 5′ upstream sequences, can affect initiation and thus translation frequencies, the 5′ noncoding region from the M1 and S4 genes were inserted adjacent to the initiating AUG_{14} of the S1 gene that has a relatively low translation efficiency. Surprisingly, both these constructs yielded similar, fourfold enhancement in σ1 protein synthesis in reticulocyte extract, indicating that the nature of the 5′ noncoding sequence in the M1 gene is not responsible for poor translatability (RONER et al. 1989).

5.2 Detection Using Trp-E-μ2 Fusion Protein-Induced Antibody

Since anti-reovirus antisera have minimal or undetectable reactivity with μ2 protein and anti-μ2 monoclonal antibodies are not available, μ2-specific antibody was induced using recombinant μ2 protein synthesized in *Escherichia coli.* T1 and T3 μ2-specific antibody produced through immunization of rabbits with Trp-E-μ2 fusion proteins cross-reacts strongly with each alternate type of μ2 protein (ZOU and BROWN 1996a), which is not surprising, since the μ2 proteins of T1 and T3 have 98.6% amino acid homology (ZOU and BROWN 1992a). These antibodies are effective for radioimmunoprecipitation (RIPA) of μ2 proteins from in vitro translation products of M1 cDNA transcripts and melted dsRNA of reoviruses that were translated using rabbit reticulocyte lysate (Fig. 1). The first five lanes show the proteins in T1-infected L cell lysates that were sequentially immunoprecipitated with anti-T1 reovirus serum and anti-μ2 fusion protein in both possible orders of reaction. The amount of μ2 protein is similar whether it follows or precedes subsequent immunoprecipitation of the same lysate with anti-T1 immune serum, as predicted by its lack of reactivity with μ2 protein. The proteins in the lane "2nd precip anti-μ2" are likely due to residual protein A beads from the first immuno-precipitation with the anti-virus antibody, since the precipitation with the anti-μ2 antibody without prior precipitation with anti-reovirus antibody (lane "anti-μ2") did not have these protein bands. It can also be seen from this sequential precip-itation that the concentration of μ2 protein in infected cells was relatively high when compared with the abundant proteins μ1c and σ3. I am currently reassessing μ2 protein levels during infection where T1 Lang induces much higher levels of μ2 than T3 Dearing (Fig. 1; J.L. MBISA and E.G. BROWN, unpublished). In vitro translation of melted dsRNA yields relatively high levels of μ2 protein that are comparable to the levels of the other viral proteins (Fig. 1). This is in contrast to the low levels of in vivo translation seen on constitutive expression of recombinant μ2 protein that will be described in detail later in this review. RIPA of in vitro translation products gave other smaller bands that represented presumably short translation products from the same reading frame as the full-size μ2 protein due to internal initiation or premature termination. These results indicated that the antibody induced in rabbits by μ2 fusion proteins expressed in *E. coli* reacted specifically with reovirus μ2 proteins.

5.3 Dependence of Constitutive Expression on Host Cell Type and Serotype

Characterization of reovirus M1 deletion mutants has identified the consensus termini of M1 required for replication and encapsidation. This information could be applicable to production of a packaging and selection system for the introduction of an M1 analogue containing the identified consensus sequences into reovirus. However, reovirus possessing an M1 analogue in place of the wild-type M1 will be replication defective. An approach taken in my laboratory to complement defective M1 genes was to develop stable cell lines expressing the μ2 protein, the M1 gene product.

M1-containing monocistronic or dicistronic DNA constructs were generated for the expression of μ2 proteins in transfected cells (Zou and Brown 1996a). T1 M1 and T3 M1 gene cDNA sequences were subcloned into a mammalian expression vector pKJ1 under the control of the mouse phosphoplycerate kinase (pgk) promoter (Boer et al. 1990; McBurney et al. 1991). The pgk promoter drives the normal house-keeping gene, phosphoglycerate kinase, and is thus a promising promoter for constitutive expression. The neomycin-resistant gene from transposon Tn5 (Beck et al. 1982) was used as a selectable marker for the selection of transfectants. Three types of M1 gene-expressing constructs have been used for μ2 expression (Fig. 2). For generation of monocistronic constructs (pKJ1M1neo), neo plus the SV40 promoter and the polyA signal were inserted into pKJ1 downstream of the pgk polyadenylation signal. The dicistronic constructs (M1CN) were generated by insertion of neo into pKJ1 followed by insertion of M1 upstream of the neo gene. The neo gene was translated due to the addition of a cap-independent translation initiation element (CITE) from encephalomyocarditis virus (EMCV), pCITE-1 (Novagene), between M1 and neo such that neo was fused with the two in-frame AUG codons. Further modification was the inclusion of the bovine papillomavirus (BPV) transforming fragment (69% portion) that increases expression efficiency through episomal replication of the transfecting plasmid and thus provides amplification of the vectored gene (reviewed by DiMaio 1987).

Mouse cell lines L929, NIH/3T3, and C127I, transfected with specific DNA constructs and selected with G 418, resulted in stable transfectants. Transfection of mouse fibroblast L929 cells with the T1 and T3 μ2 monocistronic construct pKJ1M1neo gave low, but detectable levels of T1μ2 and possibly T3μ2 expression in the uncloned pools of G418-resistant transfectants. However, screening of 18 clones of cells transfected by T1 M1 and 15 clones of cells transfected by T3 M1 did not give any clones that expressed a detectable level of μ2 protein. This illustrates the inherently low translation efficiency of both T1 and T3 M1 genes. Expression of the neo gene via the pgk promoter produces easily detectable neo protein. Flanking the neo gene by the conserved terminal 135-nucleotide 5′ and 232-nucleotide 3′ regions from the M1 gene gave similar levels of a M1-neo fusion protein, indicating that the reduced translation efficiency of the M1 gene is not due to terminal regions that include the noncoding regions but is rather a function of the body of the M1 gene (Zou and Brown 1996a).

In the dicistronic construct M1CN, both M1 and neo were put into one construct under one promoter, with M1 upstream of neo so that selection with G418 would select for higher levels of μ2 expression. Dicistronic constructs give higher μ2 expression, with T1 μ2 much higher than T3 μ2, which is undetectable in the exposure shown here (Fig. 2), indicating the serotype dependence of μ2 expression. The highest level of μ2 expression was seen for clone T1-11-1, which was 5% of the level seen in infected cells. The half-life of T1 μ2 protein produced in T1-11-1 cells was similar to that of μ2 protein synthesized in T1 infection (approximately 6 h),

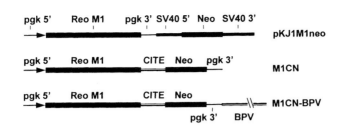

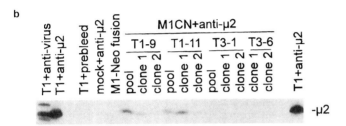

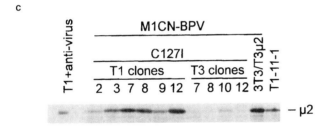

Fig. 2. a Structure of the M1 gene-containing cDNA constructs used for transfection. The mono-cistronic pKJ1M1neo contains M1 driven by the pgk promoter and neo driven by the SV40 promoter. The dicistronic M1CN has M1 and neo under the same pgk promoter with the internal ribosome entry site from encephalomyocarditis virus (EMCV), the cap-independent translation initiation element (*CITE*), between M1 and neo. M1CN-BPV is M1CN plus the 69% transforming fragment of bovine papillomavirus (*BPV*). **b** μ2 expression in L929 cells transfected with the dicistronic constructs M1CN. T1-9 and T1-11 are clones of M1CN T1M1 constructs, and T3-1 and T3-6 are clones of M1CN T3M1 constructs. The pool of transfectants from T3-6 was lost during passage. **c** μ2 expression in NIH/3T3 and C127I transfected with M1CN-BPV constructs containing T1 and T3 M1 genes. Cloned cells of trans-fectants were labeled, lysed, precipitated by anti-μ2 antibody and analyzed by 7.5% sodium dodecyl sulfate-polyacrylamide gel electrophoresis (SDS-PAGE). The 3T3 clone expressing the highest level of T3μ2 on transfection with the M1CN-BPV construct (T3-15-33 clone 1) and the L929 clone expressing the highest level of T1μ2 on transfection with the M1CN construct (T1-11-1) are shown. (Modified from Zou and Brown 1996a)

indicating that the low level of µ2 expression resulting from transfection was not due to protein instability.

The µ2 protein expression level was stable in the µ2-expressing cell lines for over 2 years (in the absence of G418). The µ2 protein from transfected cells was indistinguishable from viral µ2 by peptide mapping, indicating that the µ2 protein stably expressed by transfection was the same as authentic µ2 protein.

To increase the level of expression, the transforming fragment of BPV was introduced into the M1CN constructs (M1CN-BPV) for transfection into L929, NIH/3T3, and C127I cells. It has been shown that, in 3T3 or C127 cells, BPV-containing constructs tend to exist as plasmids with multiple copies, whereas in some other cells they tend to become integrated and consequently give low expression, as was seen for our BPV constructs in L929 cells, since expression in L929 cells was not improved (DiMaio et al. 1982; Sambrook et al. 1985). Transfections of C127I cells yielded higher T1 µ2 expression compared to T1-11-1, the L929 cell line expressing the highest level of T1 µ2 with construct M1CN (Fig. 2). The level of T3 µ2 expression was considerably lower than T1 µ2 in C127I cells, whereas expression was similar to T1 µ2 in 3T3 cells (data not shown); one clone of 3T3 transfected by T3M1-containing M1CN-BPV (lane "3T3/T3µ2" in Fig. 2C) was among the highest levels attained. These data indicate not only that it was possible to obtain higher expression of µ2, but also that µ2 protein expression was both serotype dependent and host cell restricted. In summary, through the use of different expression vectors, T1 µ2 protein was expressed to much higher levels than T3 µ2 protein in L929 and C127I cells, whereas T3 µ2 was expressed to comparable levels in NIH/3T3 cells. However, the level of µ2 expression from the BPV-containing constructs was less stable and decreased to lower but detectable levels after continuous passage of the cell lines. The BPV-containing plasmid presumably decreased in copy number or rearranged during serial passage, as documented for this type of vector (DiMaio 1987; Kitamura et al. 1991). These data strongly suggest that µ2 protein expression is modified by interaction with host factors.

6 Constitutive µ2 Expression Complements the Growth of the M1 Temperature-Sensitive Mutant *tsH11.2*

One of the goals of expressing µ2 stably in mammalian cells is the complementation of reoviruses containing a foreign gene in place of authentic M1 and thus lacking functional µ2 protein. To assess the ability of µ2 expression to complement a *ts* defect, the µ2-expressing L929 clone T1-11-1 was infected with the M1 *ts* mutant *tsH11.2*, which is restricted in replication at temperatures higher than 39 °C. The plaque-forming efficiency of *tsH11.2* on µ2-expressing cells was over 200 times higher than that from normal L929 at 39.5 °C (6.2×10^7 PFU/ml versus

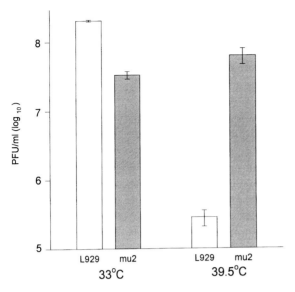

Fig. 3. Plaque formation of the M1 temperature-sensitive (*ts*) mutant *tsH11.2* at permissive (33 °C) and nonpermissive temperature (39.5 °C) on normal L929 and the µ2-expressing T1-11-1 cells. *PFU*, plaque-forming units. (From Zou and Brown 1996a)

2.8×10^5 PFU/ml), indicating that the µ2 protein constitutively expressed in L929 cells was able to complement the replication of the M1 *ts* mutant *tsH11.2* (Fig. 3).

The complementation is specific to the M1 gene defect, since infection of µ2-expressing T1-11-1 cells, with three *ts* mutants, *tsH11.2*, *tsC447*, and *tsG453*, that have *ts* mutations mapped to segments M1, S2, and S4, respectively, only rescued *tsH11.2* replication (Zou and Brown 1996a).

To confirm that the complementation was phenotypic and not the result of genetic reversions or acquisition of the M1 gene expressed in T1-11-1 cells, the virus harvests of *tsH11.2* from T1-11-1 cells at 39.5 °C were also titrated at 39.5 °C. The titers for *tsH11.2* at 39.5 °C versus 33 °C were 4.5×10^4 and 2.9×10^7 PFU/ml, respectively, indicating that the complementation was indeed phenotypic, since the efficiency of plating of *tsH11.2* was similar whether grown on wild-type L929 cells (1.5×10^{-3}) or on µ2-expressing T1-11-1 cells (1.4×10^{-3}).

Time course studies showed that, in µ2-expressing L929 cells (T1-11-1), the *ts* mutant grew faster and reached peak titers 6–12 h earlier than in normal L929 at 39.5 °C. Furthermore, in normal L929 at nonpermissive temperature, the virus titer decreased after reaching its peak at 24 h postinfection, suggesting that *ts* virus produced in wild-type L929 cells at nonpermissive temperature was not as stable as virus produced in µ2-expressing L929 cells (Zou and Brown 1996a).

The complementation of *tsH11.2* indicates that the µ2 protein expressed from transfected M1 DNA constructs is functional and that the expression level reached in T1-11-1 is sufficient to support reovirus replication. Therefore, these L929 cell lines that constitutively and stably express μ2 protein may provide a means to cultivate engineered reoviruses containing foreign genes. Initial attempts to isolate M1 deletion mutants on μ2-expressing T1-11-1 cells by screening 800 plaques from M1 deletion-containing stocks for plaque formation on μ2-expressing, but not wild-type L929 cells have been unsuccessful (B. LASIA and E.G. BROWN, unpublished). It is possible that one or more functions of μ2 protein require higher levels of μ2 protein to be complemented. Initial attempts to isolate M1 deletion mutants on L cells expressing high levels of μ2 protein via BPV-containing expression vectors indicate that high μ2 expression is inhibitory to plaque formation (E.G. BROWN, unpublished). This suggests that preexisting high levels of μ2 protein inhibit virus replication and possibly that μ2 expression must be coordinated with specific stages in replication to complement M1 deletion mutants.

As for the mechanism of complementation, the wild-type μ2 protein expressed in the transfectants presumably supplied the missing function of the mutant *tsH11.2*. Given that about 12 copies of μ2 protein are present in the interior of the virion core, it is assumed that this virion-associated protein supplies at least one critical, probably enzymatic function associated with RNA synthesis. If this function preceded or involved transcription, then complementation of that defect would not be possible unless the virion was functionally wild type but genetically mutant, as is *tsH11.2* grown at permissive temperature. The μ2 protein is present at relatively high concentrations in infected cells, relative to other abundant proteins μ1c and σ3, and may serve other functions before or concomitant with virion assembly, where the μ2 protein is sequestered in the core of the virion. Without knowing the functional and structural roles of μ2 protein, the mechanism for complementation cannot be predicted. As the μ2 protein from *tsH11.2* could not be differentiated from the wild-type μ2 either by electrophoresis or by antibody detection (data not shown), we were not able to show whether the wild-type μ2 was incorporated into the virion or whether it merely functioned in *trans*. The reduced stability of *tsH11.2* produced at nonpermissive temperature in wild-type L929 relative to μ2-expressing T1-11-1 cells suggests that viruses incorporated the wild-type μ2 protein from the host cells in order to be more stable. Further work on the nature of complementation by μ2 expression may shed light on the role it plays in specific stages in reovirus infection.

The μ2-expressing cell lines can also be used to study the function of μ2 and the interactions of μ2 with host cells. T1 and T3 μ2 expression levels differed significantly in the same type of cells and varied among different cell types. These data suggest interactions either between host factors and the M1 mRNA or between host factors and the μ2 protein. Apparently, either the 51 nucleotide substitutions and/or the ten amino acid differences between T1 and T3 (ZOU and BROWN 1992a) are responsible for the differences in μ2 expression. It is possible that structural differences between T1 Lang and T3 Dearing M1 genes control host-dependent efficiency of expression or, alternatively, that these μ2 proteins differ in host cell-dependent toxicity, which can select against high T3 μ2 expression.

7 Initiation of Translation of the M1 Gene from the First AUG at Nucleotide Position 14

By transfection of L929 cells with M1-containing dicistronic DNA constructs driven by the mouse pgk promoter, we obtained stable cell lines which express μ2 protein at 5% of the level of reovirus-infected cells (Zou and Brown 1996a). Although this level of expression was later shown to be sufficient to complement the growth of a reovirus *ts* strain with a M1 gene defect, efforts were made to try to increase the μ2 protein expression level. For both T3 Dearing and T1 Lang, the M1 gene is 2304 nucleotides in length with a large open reading frame extending from nucleotide 14 to 2224 with two in-frame translation initiation codons starting at nucleotides 14 (AUG_{14}) and 161 (AUG_{161}), respectively (Wiener et al. 1989; Zou and Brown 1992a). It was reported that in vitro translation of reovirus M1 can initiate at either of the two in-frame initiation codons, but in reovirus-infected cells only AUG_{161} is used (Roner et al. 1993).

The M1-containing dicistronic DNA constructs were first modified by deletion of the 5′ terminal 144 or 160 nucleotides with or without changes in the AUG_{161} sequence context, followed by removal of the 3′ untranslated region (UTR). It was found that deletion of the 80-nucleotide 3′ UTR did not improve μ2 expression and that deletion of the 144 or 160 nucleotides of the 5′ terminal sequence abolished constitutive expression (Zou and Brown 1996b). Insertion of the T7 promoter upstream of this series of modified M1 DNA constructs for transient expression driven by the recombinant vaccinia-induced T7 polymerase show that the protein products from constructs with 5′ terminal 144 or 160 nucleotides deleted were smaller than the μ2 protein expressed from the full-size M1 constructs, whereas the latter was the same size as μ2 protein precipitated from virus-infected cells by anti-μ2 antibody (Fig. 4). These data demonstrate that the translation of reovirus M1 gene initiates from the first AUG in both infected and transfected cells. The two constructs with the 5′ terminal 144 nucleotides deleted, but without the context of AUG_{161} changed, gave the lowest level of expression, which is not surprising since this AUG is in a very poor context for initiation, with U at −3 and +4 positions. Mutations that produced a more favorable Kozak consensus sequence (Kozak 1987), with A at the −3 and +4 positions, gave much higher levels of the smaller protein product. Furthermore, the protein band from the construct pT7M1CNk (Fig. 4) was also smaller than the authentic μ2 proteins and was apparently the same size as those from other 5′ end-truncated constructs. Because the translation initiation with this construct was from the EMCV AUG overlapping with the M1 AUG_{161}, the size of the protein product from this construct demonstrated that the smaller-size product of 68-kDa (compared to 73 kDa for full-length μ2 protein) was not the result of translation initiation from an AUG codon further downstream at position 332 (weak Kozak sequence) or 443 (good Kozak sequence ACAAUGA); therefore, the smaller protein band expressed from all the 5′ end-truncated M1 constructs was indeed translated though initiation at AUG_{161} of the M1 gene. Thus translation of μ2 protein follows the first AUG rule, whereby AUG_{14}, which

a

pGEM7 constructs

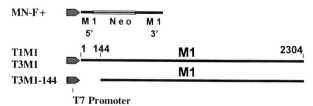

pT7M1CN

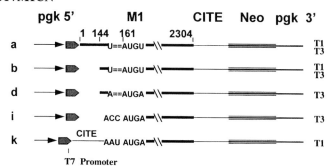

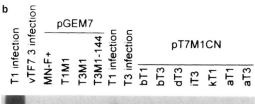

b

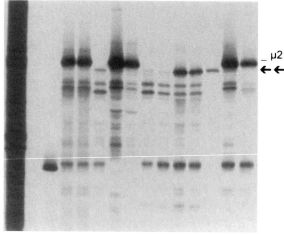

Fig. 4a, b. Expression of two different-sized proteins from intact or 5′-end-truncated M1 constructs. **a** Constructs used for transfection. **b** L929 cells were transfected followed by infection with the recombinant vaccinia vTF7.3. Cell lysates were extracted and tested by radioimmunoprecipitation (RIPA) using the anti-μ2 antibody. *MN-F+*, Neo gene fused to the M1 5′-terminal 135 nucleotides; *T3M1-144*, the T3M1 5′-terminal 144 nucleotides deleted; *b*, the M1 5′-terminal 144 nucleotides deleted; *d*, deletion of the 144 nucleotides plus mutations of the −3 and +4 positions; *i*, deletion of 160 nucleotides with a +4 mutation; *k*, cap-independent translation initiation element (CITE) in place of the 5′ 160 nucleotides; *a*, full-length M1. The *double arrow* indicates the smaller protein from the truncated M1 gene. (From ZOU and BROWN 1996b)

possesses an excellent Kozak context, is utilized in both infected and transfected cells.

In attempts to improve the expression of μ2 protein, RONER et al. (1993) found that addition of short, unrelated sequences of about 20 nucleotides to the 5′ or 3′ end of the full-length M1 message resulted in a second, larger protein in addition to authentic μ2 protein on in vitro translation in reticulocyte extracts. This observation was interpreted as a choice between initiation at AUG_{14} and AUG_{161} in the +1 reading frame on the basis of translation of a construct that deleted the first 15 nucleotides of the M1 gene, altering AUG_{14} to produce CCG. Since in vitro translation of this construct only produced protein indistinguishable from μ2 protein, they concluded that translation of M1 initiates from AUG_{161}. They did not test their hypothesis directly by deleting portions of the gene between positions 14 and 160. When I did this, a truncated protein was synthesized, since this region codes for the amino-terminal 49 amino acids of μ2 protein. The production of a larger protein from the M1 gene appears to be an in vitro artifact resulting from translation of modified M1 mRNA in reticulocyte extract, since these constructs did not induce larger proteins when expressed in vaccinia virus (RONER et al. 1993; ZOU and BROWN 1996b).

Acknowledgements. Research in my laboratory has been funded by the Natural Sciences and Engineering Research Council of Canada.

References

Acs G, Schonberg M, Christman J, Levin DH, Silverstein SC (1971) Mechanism of reovirus double-stranded ribonucleic acid synthesis in vivo and in vitro. J Virol 8:684–689
Ahmed R, Graham AF (1977) Persistent infections in L cells with temperature-sensitive mutants of reovirus. J Virol 23:250–262
Ahmed R, Chakraborty PR, Fields BN (1980a) Genetic variation during lytic virus infection: high passage stocks of reovirus contain temperature-sensitive mutants. J Virol 34:285–287
Ahmed R, Chakraborty PR, Graham AF, Ramig RF, Fields BN (1980b) Genetic variation during persistent reovirus infection: presence of extragenically suppressed temperature-sensitive lesions in wild-type virus isolated from persistently infected cells. J Virol 34:383–389
Antczak JB, Chmelo R, Pickup DJ, Joklik WK (1982) Sequences at both termini of the ten genes of reovirus serotype 3 (strain Dearing). Virol 121:307–319

Anzola JV, Xu Z, Asamizu T, Nuss DL (1987) Segment-specific inverted repeats found adjacent to conserved terminal sequences in wound tumor virus genome and defective interfering RNAs. Proc Natl Acad Sci USA 84:8301–8305

Beck E, Ludwig EA, Reiss B, Schaller H (1982) Nucleotide sequence and exact location of the neomycin phosphotransferase gene. Gene 19:327–336

Boer PH, Potten H, Adra CN, Jardine K, Mullhofer G, McBurney, MW (1990) Polymorphisms in the coding and noncoding regions of murine Pgk-1 alleles. Biochem Genet 28:299–307

Both GW, Lavi S, Shatkin AJ (1975) Synthesis of all the gene products of the reovirus genome in vivo and in vitro. Cell 4:173–180

Brown EG, Nibert ML, Fields BN (1983) The L2 gene of reovirus serotype 3 controls the capacity to interfere, accumulate deletions and establish persistent infection. In: Compans RW, Bishop DHL (eds) Double-stranded RNA viruses. Elsevier Science, Amsterdam

Coombs KM (1996) Identification and characterization of a double-stranded RNA reovirus temperature-sensitive mutant defective in minor core protein μ2. J Virol 70:4237–4245

Dall DJ, Anzola JV, Xu Z, Nuss DL (1990) Structure-specific binding of wound tumor virus transcripts by a host factor: involvement of both terminal nucleotide domains. Virology 179:599–608

DiMaio D (1987) Papillomavirus cloning vectors. In: Salzman NP, Howley PM (eds) The Papovaviridae. 2. The papillomaviruses. Plenum, New York

DiMaio D, Treisman R, Maniatis T (1982) Bovine papillomavirus vector that propagates as a plasmid in both mouse and bacterial cells. Proc Natl Acad Sci USA 79:4030–4034

Gaillard RK, Joklik WK (1985) The relative translation efficiencies of reovirus messenger RNAs. Virology 147:336–348

Gaillard RK, Li JK-K, Keene JD, Joklik WK (1982) The sequences at the termini of four genes of the three reovirus serotypes. Virology 121:320–326

Haller B, Barkon ML, Vogler GP, Virgin IV HW (1995) Genetic mapping of reovirus virulence and organ tropism in severe combined immunodeficient mice: organ-specific virulence genes. J Virol 69:357–364

Joklik WK (1974) Reproduction of the reoviridae. In: Fraenkel-Conrat H, Wagner RR (eds) Comprehensive virology, vol 2. Plenum, New York

Kavenoff R, Talcove D, Mudd JA (1975) Genome-sized RNA from reovirus particles. Proc Natl Acad Sci USA 72:4317–4321

Kitamura Y, Naito A, Yoshikura H (1991) Illegitimate recombination in a bovine papillomavirus shuttle vector: a high level of site specificity. Biochem Biophys Res Commun 179:251–258

Kozak M (1987) An analysis of 5′-noncoding sequences from 699 vertebrate messenger RNAs. Nucleic Acids Res 15:8125–8148

Loh PC, Shatkin AJ (1968) Structural proteins of reovirus. J Virol 2:1353–1359

Matoba Y, Sherry B, Fields BN, Smith TW (1991) Identification of the viral genes responsible for growth of strains of reovirus in cultured mouse heart cells. J Clin Invest 87:1628–1633

Matoba Y, Colucci WS, Fields BN, Smith TW (1993) The reovirus M1 gene determines the relative capacity of growth of reovirus in cultured bovine aortic endothelial cells. J Clin Invest 92:2883–2888

McBurney MW, Sutherland LC, Adra CN, Leclair B, Rudnicki MA, Jardine K (1991) The mouse Pgk-1 gene promoter contains an upstream activator sequence. Nucleic Acids Res 19:5755–5761

McCrae MA, Joklik WK (1978) The nature of the polypeptide encoded by each of the double-stranded RNA segments of reovirus type 3. Virology 89:578–593

Millward S, Graham AF (1970) Structural studies on reovirus: discontinuities in the genome. Proc Natl Acad Sci USA 65:422–429

Moody MD, Joklik WK (1989) The function of reovirus proteins during the reovirus multiplication cycle: analysis using monoreassortants. Virology 173:437–446

Nibert ML, Schiff LA, Fields BN (1996) Reoviruses and their replication In: Fields BN, Knipe DM, Howley PM (eds) Fundamental virology, 3rd edn. Lippincott-Raven, Philadelphia

Nuss DL (1988) Deletion mutants of double-stranded RNA genetic elements found in plants and fungi. In: Domingo E, Holland JJ, Alquist P (eds) RNA genetics, vol 2. CRC, Boca Raton

Nuss DL, Dall DJ (1990) Structural and functional properties of plant reovirus genomes. Adv Virus Res 38:272–276

Nuss DL, Summers D (1984) Variant dsRNAs associated with transmission-defective isolates of wound tumor virus represent terminally conserved remnants of genome segments. Virology 133:276–288

Perrault J (1981) Origin and replication of defective interfering particles. Curr Top Microbiol Immunol 93:151–207

Roner MR, Gaillard RK, Joklik WK (1989) Control of reovirus messenger RNA translation efficiency by the regions upstream of initiation codons. Virology 168:292–301

Sambrook J, Rodgers L, White J, Gething M-J (1985) Lines of BPV-transformed murine cells that constitutively express influenza virus hemagglutinin. EMBO J 4:91–103

Schlesinger S (1988) The generation and amplification of defective interfering RNAs. In: Domingo E, Holland JJ, Alquist P (eds) RNA genetics, vol 2. CRC, Boca Raton

Sherry B, Blum MA (1994) Multiple viral core proteins are determinants of reovirus-induced acute myocarditis. J Virol 68:8461–8465

Sherry B, Fields BN (1989) The reovirus M1 gene, encoding a viral core protein, is associated with the myocarditic phenotype of a reovirus variant. J Virol 63:4850–4856

Smith RE, Zweerink HJ, Joklik WK (1969) Polypeptide components of virions, top component and cores of reovirus type 3. Virology 39:791–810

Spendlove RS, McLain ME, Lennette EH (1970) Enhancement of reovirus infectivity by extracellular removal or alteration of the virus capsid by proteolytic enzymes. J Gen Virol 8:83–94

Sukuma S, Watanabe Y (1972) Reovirus replicase-directed synthesis of double stranded ribonucleic acid. J Virol 10:628–638

Theile DJ, Hannig EM, Leibowitz J (1984) Genome structure and expression of a defective interfering mutant of the killer virus of yeast. Virology 137:20–31

Wiener JR, Bartlett JA, Joklik WK (1989) The sequences of reovirus serotype 3 genome segments M1 and M3 encoding the minor protein $\mu2$ and the major nonstructural protein μNS, respectively. Virology 169:293–304

Xu Z, Anzola JV, Nalin CM, Nuss DL (1989) The 3′-terminal sequence of a wound tumor virus transcript can influence conformational and functional properties associated with the 5′-terminus. Virology 170:511–522

Yin P, Cheang M, Coombs KM (1996) The M1 gene is associated with differences in the temperature optimum of the transcriptase activity. J Virol 70:1223–1227

Zarbl H, Millward S (1983) The reovirus multiplication cycle. In: Joklik WK (ed) The reoviridae. Plenum, New York

Zou S, Brown EG (1992a) Nucleotide sequence comparison of the M1 genome segment of reovirus type 1 Lang and type 3 Dearing. Virus Res 22:159–164

Zou S, Brown EG (1992b) Identification of sequence elements containing signals for replication and encapsidation of the reovirus M1 genome segment. Virology 188:377–388

Zou S, Brown EG (1996a) Stable expression of the reovirus $\mu2$ protein in mouse L cells complements the growth of a reovirus ts mutant with a defect in its M1 gene. Virology 217:42–48

Zou S, Brown EG (1996b) Translation of the reovirus M1 gene initiates from the first AUG codon in both infected and transfected cells. Virus Res 40:75–89

Zucker M, Stiegler P (1981) Optimal computer folding of large RNA sequences using thermodynamics and auxiliary information. Nucleic Acids Res 9:133–148

Zweerink J, Ito Y, Matsuhisa T (1972) Synthesis of reovirus double-stranded RNA within virion-like particles. Virology 50:349–358

Subject Index

Subject Index of Companion Volume 233/II

Printing: Saladruck, Berlin
Binding: Buchbinderei Lüderitz & Bauer, Berlin

Current Topics in Microbiology and Immunology

Volumes published since 1989 (and still available)

Vol. 212: **Vainio, Olli; Imhof, Beat A. (Eds.):** Immunology and Developmental Biology of the Chicken. 1996. 43 figs. IX, 281 pp. ISBN 3-540-60585-1

Vol. 213/I: **Günthert, Ursula; Birchmeier, Walter (Eds.):** Attempts to Understand Metastasis Formation I. 1996. 35 figs. XV, 293 pp. ISBN 3-540-60680-7

Vol. 213/II: **Günthert, Ursula; Birchmeier, Walter (Eds.):** Attempts to Understand Metastasis Formation II. 1996. 33 figs. XV, 288 pp. ISBN 3-540-60681-5

Vol. 213/III: **Günthert, Ursula; Schlag, Peter M.; Birchmeier, Walter (Eds.):** Attempts to Understand Metastasis Formation III. 1996. 14 figs. XV, 262 pp. ISBN 3-540-60682-3

Vol. 214: **Kräusslich, Hans-Georg (Ed.):** Morphogenesis and Maturation of Retroviruses. 1996. 34 figs. XI, 344 pp. ISBN 3-540-60928-8

Vol. 215: **Shinnick, Thomas M. (Ed.):** Tuberculosis. 1996. 46 figs. XI, 307 pp. ISBN 3-540-60985-7

Vol. 216: **Rietschel, Ernst Th.; Wagner, Hermann (Eds.):** Pathology of Septic Shock. 1996. 34 figs. X, 321 pp. ISBN 3-540-61026-X

Vol. 217: **Jessberger, Rolf; Lieber, Michael R. (Eds.):** Molecular Analysis of DNA Rearrangements in the Immune System. 1996. 43 figs. IX, 224 pp. ISBN 3-540-61037-5

Vol. 218: **Berns, Kenneth I.; Giraud, Catherine (Eds.):** Adeno-Associated Virus (AAV) Vectors in Gene Therapy. 1996. 38 figs. IX,173 pp. ISBN 3-540-61076-6

Vol. 219: **Gross, Uwe (Ed.):** Toxoplasma gondii. 1996. 31 figs. XI, 274 pp. ISBN 3-540-61300-5

Vol. 220: **Rauscher, Frank J. III; Vogt, Peter K. (Eds.):** Chromosomal Translocations and Oncogenic Transcription Factors. 1997. 28 figs. XI, 166 pp. ISBN 3-540-61402-8

Vol. 221: **Kastan, Michael B. (Ed.):** Genetic Instability and Tumorigenesis. 1997. 12 figs.VII, 180 pp. ISBN 3-540-61518-0

Vol. 222: **Olding, Lars B. (Ed.):** Reproductive Immunology. 1997. 17 figs. XII, 219 pp. ISBN 3-540-61888-0

Vol. 223: **Tracy, S.; Chapman, N. M.; Mahy, B. W. J. (Eds.):** The Coxsackie B Viruses. 1997. 37 figs. VIII, 336 pp. ISBN 3-540-62390-6

Vol. 224: **Potter, Michael; Melchers, Fritz (Eds.):** C-Myc in B-Cell Neoplasia. 1997. 94 figs. XII, 291 pp. ISBN 3-540-62892-4

Vol. 225: **Vogt, Peter K.; Mahan, Michael J. (Eds.):** Bacterial Infection: Close Encounters at the Host Pathogen Interface. 1998. 15 figs. IX, 169 pp. ISBN 3-540-63260-3

Vol. 226: **Koprowski, Hilary; Weiner, David B. (Eds.):** DNA Vaccination/Genetic Vaccination. 1998. 31 figs. XVIII, 198 pp. ISBN 3-540-63392-8

Vol. 227: **Vogt, Peter K.; Reed, Steven I. (Eds.):** Cyclin Dependent Kinase (CDK) Inhibitors. 1998. 15 figs. XII, 169 pp. ISBN 3-540-63429-0

Vol. 228: **Pawson, Anthony I. (Ed.):** Protein Modules in Signal Transduction. 1998. 42 figs. IX, 368 pp. ISBN 3-540-63396-0

Vol. 229: **Kelsoe, Garnett; Flajnik, Martin (Eds.):** Somatic Diversification of Immune Responses. 1998. 38 figs. IX, 221 pp. ISBN 3-540-63608-0

Vol. 230: **Kärre, Klas; Colonna, Marco (Eds.):** Specificity, Function, and Development of NK Cells. 1998. 22 figs. IX, 248 pp. ISBN 3-540-63941-1

Vol. 231: **Holzmann, Bernhard; Wagner, Hermann (Eds.):** Leukocyte Integrins in the Immune System and Malignant Disease. 1998. 40 figs. XIII, 189 pp. ISBN 3-540-63609-9

Vol. 232: **Whitton, J. Lindsay (Ed.):** Antigen Presentation. 1998. 11 figs. IX, 244 pp. ISBN 3-540-63813-X

Vol. 233/II: **Tyler, Kenneth L.; Oldstone, Michael B. A. (Eds.):** Reoviruses II. 1998. 44 figs. XV, 187 pp. ISBN 3-540-63947-0